W9-BFL-500

WITHDRAWN

Shawnee Books

Also in this series

THE NEXT NEW MADRID
EARTHQUAKE

A Survival Guide for the Midwest

WILLIAM ATKINSON

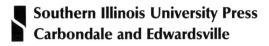
Southern Illinois University Press
Carbondale and Edwardsville

Copyright © 1989 by Board of Trustees, Southern Illinois University
All rights reserved
Printed in the United States of America
Edited by Curtis L. Clark
Designed by Loretta Vincent
Production supervised by Natalia Nadraga

92 91 90 89   5 4 3 2 1

Library of Congress Cataloging-in-Publication Data

Atkinson, William, 1951–
   The next New Madrid earthquake.

   (Shawnee books)
   Bibliography: p.
   Includes index.
   1. Earthquakes—Missouri—New Madrid Region.
   2. Disaster relief—Missouri—New Madrid Region.
   I. Title.
QE535.2.U6A85 1988 363.3′495′09778895 86-31637
ISBN 0-8093-1319-7
ISBN 0-8093-1320-0 (pbk.)

The paper used in this publication meets the minimum requirements
of American National Standard for Information Sciences—Permanence
of Paper for Printed Library Materials, ANSI Z39.48-1984. ∞

Contents

Danville Public Library
Danville, Illinois

Illustrations

Maps

Charts

Tables

THE NEXT NEW MADRID EARTHQUAKE

Introduction

For anyone who has not heard about the New Madrid Seismic Zone, a month's worth of copies of almost any newspaper in the Mississippi Valley (southern Illinois, southeastern Missouri, southern Indiana, northeastern Arkansas, northwestern Mississippi, western Tennessee, and western Kentucky) should supply at least one article on the subject. Simply stated, this concern and publicity historically stem from a series of three earthquakes that, a little less than two hundred years ago (in late 1811 and early 1812), ripped the Mississippi Valley apart, being felt as far away as the Atlantic Ocean. Scientists who specialize in the study of Mississippi Valley earthquakes insist there someday will be a repeat performance of that cataclysm.

Although such an event will definitely occur *sometime*, however, scientists do not necessarily predict for the near future an earthquake as powerful as those in the 1811–1812 sequence (which were the most powerful ever felt on the North American continent). Scientists do say, though, that the area is overdue for a *moderately* powerful tremor—one which will still cause major damage and undoubtedly some casualties. Moderately strong earthquakes, scientists speculate, occur approximately every seventy-five to one hundred years in the region,[1] and the last one occurred in 1895. If ninety years is used as a measure, the region has been living on borrowed time since 1985. With each passing year, the chances of the earthquake increase, and with each passing year the energy building up underneath our feet increases. So with each passing year, the inevitable earthquake is becoming more powerful, while the state of readiness in the Mississippi Valley remains woefully inadequate.

The different segments of the news media have been among the few entities in the region to notice the probability of dire circumstances and to attempt to inform the public. There has been, is, and will be only so much the news media can do, however. While they have been able to do an excellent job of reminding the region's residents of the coming danger with regular stories on the subject, they are not equipped to develop, organize, and disseminate a comprehensive survey, examination, and analysis of the problem. The problem is simply too large in scope.

Similarly, regional scientists who study the problem have done an excellent job of providing valuable information to the media so the information can be disseminated to the public and to local officials. Many of these scientists have been working on the problem for years, and only through their dedication and persistence have people in the region finally begun to sit up and take notice. Again, however, there is only so much the region's scientists can do.

The branches and agencies of the federal government have also addressed the problem, and through their coordinated efforts have taken steps to both assess the problem and to develop solutions. They have been instrumental in developing preparedness and mitigation programs that can be adapted for regional and local use. Since it is a regional problem, however, there is only so much the federal government can be expected to do.

Where, then, can one find information about major earthquakes in the Mississippi Valley and how to prepare for them? It is hoped this book will provide needed information by addressing the following areas: Emphasizing the past, chapter 1 outlines what actually happened during the 1811–1812 cataclysms—explaining how powerful the earthquakes were and what kinds of damage and destruction they caused—while chapter 2 discusses powerful earthquakes that have occurred in the region since 1812. Emphasizing the present, chapter 3 explains what earthquakes are, how they occur, and where they are likely to occur; chapter 4 examines the possibilities of earthquake prediction and prevention; and chapter 5 shows what has been done to prepare for major earthquakes in the Mississippi Valley and who has done it. Finally, emphasizing the future, chapter 6 suggests what still needs to be done to prepare for a major earthquake—and what private citizens specifically can do—while chapter 7 studies the kinds of damage and destruction that can be expected when the next major New Madrid earthquake strikes, as well as what other kinds of problems (socioeconomic, psychological, and so forth) the earthquake will cause over the long term.

This book is intended to help lay people comprehend the problem the Mississippi Valley is faced with. It will not, therefore, tell earth scientists anything they do not already know about earthquakes, though it may tell them something they do not know about the psychological effects of earthquakes. It will not tell sociologists or psychologists anything they do not already know about human response to major disaster, though it may tell them something they do not know about the mechanics of earthquakes.

This book is also specifically intended to provide valuable information to decision-makers and other professional people in the region whose careers relate in one way or another to the problem of earthquakes and the subsequent damage and chaos they leave behind. These professionals include:

● state, county, and city government officials (such as emergency service personnel, mayors, and members of city councils and county boards);
● public school personnel (superintendents, school boards, and teachers);
● private professionals (engineers, architects, business and industry executives, chamber of commerce executives, utility company executives, and others);
● media professionals (including radio, television, and newspaper executives, editors, and reporters); and

● medical and mental health care professionals (such as hospital executives and boards, doctors, nurses, emergency medical technicians, and psychologists).

With tens of thousands of lives and businesses being threatened by a powerful earthquake in the region, most people in the seven-state region would be affected, and, therefore, most people in the region would benefit by reading at least a portion of this book.

PART ONE: THE PAST

1. The Destruction of the Mississippi Valley

The "Great Comet of 1811" and an Uneasy Calm

The Azores, a group of islands in the North Atlantic off Portugal, are far enough away from the United States of America that when a submarine volcano shot up from the depths of the ocean on January 30, 1811, and formed a new island in the group, few if any residents of New Madrid, Missouri, paid much attention. Even tremors occurring in South Carolina and Georgia that month did not cause Midwestern residents much concern, nor did a quake rumbling through the West Indies in the Caribbean Sea just five months later.[1]

There were more important things happening closer to home. On September 26, 1811, General William Henry Harrison, governor of the Indiana Territory, took advantage of the temporary absence of the Indian leader, Tecumseh, and moved his eight hundred troops along the Wabash River to the Indian encampment near the Tippecanoe River. He arrived on November 6. Harrison won the battle that occurred the following day, but the Indians scattered, and the frontier was alive with concern about what the Indians would do upon Tecumseh's return.

Pirates, robbers, and counterfeiters were another problem, from Evansville, Indiana, to Natchez, Mississippi. Flatboats and other trading vessels moving up and down the Mississippi and Ohio rivers were well armed and often traveled in convoys for additional protection. Lawlessness flourished, and the criminal element ruled.[2]

The weather, too, was a concern to the inhabitants along the Mississippi River in and around New Madrid. Flooding of the river was heavy in the spring. Sickness followed. Summer storms were common, but strangely, lightning and thunder were absent from the frequent storms. (Residents did report hearing noises like thunder, though they came not from the storms, but rather from deep within the earth.) A heavy, uneasy stillness hung over the region.

There were other ill omens. Residents were surprised and baffled in the fall of 1811 when tens of thousands of squirrels inexplicably surrendered their forest homes and moved in solid phalanxes toward the Ohio River and drowned themselves.[3] The sky was not without its seeming warnings, either. A bright comet, later known as the Great Comet of 1811, became visible in September of that year as it streaked through the sky in its southwesterly path. It was notably brilliant and shed a glow over the forests. Comets, tradition warned, always brought doom and disaster with them. People around New Madrid may well have been concerned.

At the time, the region was home to a little over three thousand inhabitants, approximately eight hundred of whom lived in the village of New Madrid and two hundred of whom resided in Little Prairie (now called Caruthersville), thirty miles south of New Madrid. On Sunday evening, December 15, 1811, it is safe to say that most of these residents were concerned only with getting a good night's sleep. Some children doubtlessly lay awake with thoughts of Christmas. Some adults may well have also lain awake, wondering what had caused their horses to be so completely unmanageable that day.[4] Some residents, however, were not even trying to sleep. The French settlers were wide awake, reveling, celebrating, and dancing. The echoes of their merriment floated down to the river, where as many as forty flatboats were strung out along the dark bank.[5] The sleepy passengers were seeking mutual protection, concerned about Indian attacks and pirate raids. When their reverie was snapped at a little after

2:00 A.M. that moonless night, Indians and pirates became the least of their worries.

New Madrid, Missouri

At just past 2:00 A.M. on Monday morning, December 16, 1811, most of the residents of New Madrid, Missouri, were asleep when their reverie was suddenly broken by the sounds of their sturdy cabins creaking. The small town's citizens were thrown from their beds, which began to shake violently. The floors heaved wildly, and the walls groaned, wrenching from their foundations. Many of the cabins crumbled, falling to the ground in piles of rubble. Some caught on fire as the logs in the walls fell like toothpicks into the fireplaces that were meant to keep the residents warm that chilly night.

Most of the stupefied inhabitants ran from their falling homes into the cold night. They heard rumbling sounds, like distant thunder fast approaching, soon followed by deafening explosions sounding like large cannon. Mixed with the rumblings and sharp explosions was an earsplitting roar, accompanied by a loud hissing and a shrill whistling sound, all emanating from the ground below them.

As the cacophony of deafening sounds surrounded the terrified residents, a strong, noxious odor not unlike sulphur began to envelop them. The pungent gas seemed to be seeping from the very ground itself, spewing up through large cracks that began to appear. Breathing became difficult. The inhabitants began to gasp for air.

Flashes of bright light suddenly burst from the earth, which shook violently—both horizontally and vertically. In places, it rolled and undulated in waves. As the waves peaked, the earth burst open and spewed forth water, sand, rock, and coal, often as high as the treetops. In other places, the ground exploded vertically, not unlike volcanoes, without benefit of waves. Everywhere, the ground shook furiously from side to side. Residents were unable to stand. Many of them had experienced the tossing of ships on violent seas, but at this moment most would have found a roiling ocean more comfortable.

To add to the calamity, the atmosphere itself darkened—thick, black clouds of dust, dirt, gases, and other matter from the ground methodically blanketed the stricken village—and the shocks continued through the night, at between half-hour and one-hour intervals. The residents, Catholic and Protestant alike, knelt and prayed together.

By daylight, a total of twenty-seven shocks had occurred, six of them severe. Just before 7:00 A.M., when the shocks seemed to have subsided, the residents began to hear another rumbling. Suddenly, the ground burst forth again. Large waves undulated across the landscape, vomiting their contents skyward. Sections of earth blew straight up, accompanied again by the thunderous explosions that had shaken residents through the night. Carbonized wood was mixed with the water, sand, rock, and coal. The mixture fell to the ground from its fifteen-foot height, reduced to black dust.

The ground throughout the town began to sink, and a black liquid oozed from the earth, rising eventually to a depth of three feet. Large fissures opened, splitting the ground into hundreds of deep chasms, some of them seven feet wide and of untold depth. Some fissures ran for a few yards. Others are believed to have split the earth for several miles, running from southwest to northeast.[6] Many slammed shut as quickly as they opened, forcing up large quantities of water, mud, rotted wood, and rock. A boatload of castings, stored in the basement of one residence, dropped into the dark depths of one of the fissures that opened up under the home and was never seen again.[7]

The trees in the nearby forests waved frantically. Many were split up the middle as fissures opened beneath them. Oak trees, sometimes split to a height of forty feet, perched precariously, one half on one side of a fissure, the other on the other side. Many trees were simply uprooted and crashed earthward. Others snapped with a loud crack and fell to the ground. Thousands of trees, whole forests, fell in unison. Still other trees, those with more supple trunks and limbs, swayed like upside-down pendulums, only to become entangled in the branches of others nearby and locked together.

In the town itself, most of the houses had been thrown down in piles of rubble by the time the Mississippi River overflowed its banks and began to flood the town. The town's cemetery and its dead were exposed and carried away by the torrents of water. The townspeople huddled together where they could for warmth, comfort, and solace. Many had fainted, some due to the excitation, others due to vertigo. Even those who did not feel weak had enough presence of mind to remain seated or prostrate. Each succeeding shake knocked down everything and everyone that was standing.

The Missouri Bootheel and the Mississippi River

The residents of Little Prairie (now Caruthersville), a town situated about thirty miles south of New Madrid in what is now known as the Missouri Bootheel, felt the earthquakes just as strongly, if not more strongly, than did those of New Madrid. The epicenter of the quake, in fact, was closer to Little Prairie than to New Madrid. Forming a smaller community at the time, however, Little Prairie eyewitnesses were fewer in number.

The first shocks caused similar damage to the two towns, but the 7:00 A.M. shock destroyed almost every home in Little Prairie. The entire bank along the Mississippi River at Little Prairie collapsed and was carried away by the raging waters. A nearby bayou actually rose several feet, was drained of its water, and a new bayou was created nearby. The town itself and most of the land around it sank several feet and was inundated by the river. The town's cemetery, like that at New Madrid, was washed away.

Many of the settlement's residents had already fled during the night, and the hundred or so who remained at dawn, seeing the water cascading toward them in the early morning hours, realized the necessity of a quick flight. Many wished they had left earlier as they slogged through eight miles of waist-deep water on their way to higher ground. The emigration was complicated even more by the fact that many of the tired refugees slipped into the deep fissures in the ground that were now hidden by the murky, chilly flood waters. Eight days later, on Christmas Eve, the stragglers reached New Madrid, only thirty miles away.

While most of the inhabitants of the Missouri Bootheel lived in either New Madrid or Little Prairie, there were a number of other settlers living near or along other parts of the Mississippi River when the shocks struck on December 16. One such settler, living about twenty miles northeast of New Madrid, reported experiences similar to those of village residents—loud explosions, suffocating sulphurous vapors, undulating ground, deep fissures, falling trees, and stygian darkness. Many of the fissures ejected large quantities of sand and water before slamming shut. Thousands of willow trees snapped off at waist height "like pipe stems."[8] Ten miles south of Little Prairie, near the Pemiscot River, a woman leaving her home after the early morning shock to survey the damage found her well and smokehouse on the other side of the river.[9] A deep fissure had ripped through the ground and given the river a new course to follow. Throughout the region, forests fell like sticks.

Collapsing riverbanks were a common sight in the area on December 16, 1811. Hundreds of acres of riverbanks severed themselves from land and crashed into the raging Mississippi River. The river itself was in constant agitation, boiling, churning, and foaming. Previously clear, it turned to a rusty brown hue and became thick, as mud and other river-bottom debris was forced to the surface by the tremors. Previously bone-chilling cold, it became lukewarm. The underground tremors and the collapse of so many banks caused the river's level to rise—three or four feet in some places, possibly as much as fifteen feet in others. As the water levels rose and fell, velocities changed equally as quickly. In places, the river reached three times its normal speed.

Fissures opened up under the river just as they did on land, creating waves larger and more powerful than those found in oceans during storms. Water spouts shot up into the air to heights of ten feet or more. Giant whirlpools sucked debris downward with dizzying speed. In places, where fissures and chasms were particularly large and deep, waterfalls actually formed, more than one at least six feet high, and for a time, the river actually flowed upstream as a huge fissure sucked water into it from both directions. Branches and tree trunks from years and generations past, until now resting peacefully on the river's murky bottom, were propelled to the surface, where they were met by the freshly fallen trees from the caving riverbanks and together covered the surface with floating timber. Other logs from the river bottom were hurtled upward with such force that they were ejected from the water and hurled skyward. Many came back down with such force that they literally planted themselves in the river bottom and remained standing upright. One tree from a bank, with a trunk three feet in diameter, was catapulted fully one hundred yards into the water and landed upright in the middle of the river.[10]

Debris in the water, however, was the least of the problems boatmen faced. An unknown number of boats of all shapes and sizes capsized during the night and early morning, drowning their crews. The river was littered with smashed vessels for miles. Many casualties were the victims of falling banks that came crashing down on them during the night as they slept on their boats, which were tied to shore. Other boats, when their moorings broke loose, were destroyed in the middle of the river by choppy waves, whirlpools, and waterfalls. Crews were caught by surprise when the river began flowing upstream. Many vessels were driven backwards a mile or more by waves and ended up beached on shore or upstream in small tributaries and creeks. Others were not so fortunate; they were driven upstream in the middle of the river, and as it returned to its normal direction, were propelled back downstream at frightening speeds.

No one will ever know the death toll on the river. Boatmen and villagers alike reported seeing large numbers of bodies floating downstream, and the battered vessels along the shoreline and floating capsized down the river attested to the casualties. One huge barge, loaded with five hundred barrels of flour and other supplies, was found upside down along the riverbank by New Madrid, split from end to end. This was one of thirty such vessels in the area at the time. Only this and one other escaped total destruction on the river.[11]

From the Rockies to the Atlantic

It is known from historical accounts that the New Madrid earthquake of December 16, 1811, was felt in at least twenty-seven states (Alabama, Arkansas, Connecticut, Georgia, Illinois, Indiana, Iowa, Kansas, Kentucky, Louisiana, Maryland, Massachusetts, Michigan, Mississippi, Missouri, Nebraska, New Jersey, New York, North Carolina, Ohio, Oklahoma, Pennsylvania, South Carolina, Tennessee, Vermont, Virginia, and West Virginia), as well as in the District of Columbia.[12]

At nearby Jackson, Missouri, about fifty miles north of New Madrid, large trees snapped, fences and brick buildings were destroyed or nearly destroyed, and haze and dark clouds from the destruction shielded the sun for three days. At St. Louis, witnesses claimed that the first shock was preceded by flashes of light from the ground, followed by sounds like wind rustling through the trees. The clatter of windows, doors, and furniture was followed by a distant rumbling sound. A thick, hazy fog enveloped the city. A number of subsequent shocks were felt through the night, and the early morning shock toppled chimneys and split a number of stone houses. Domestic birds toppled from their perches, and many residents fled their homes.[13]

Along the Mississippi River in Arkansas, ruin was extensive, with forests being flattened and fissures, some as long as fifteen miles, splitting the earth.[14] The Chickasaw Bluffs, just north of Memphis, Tennessee, experienced a great deal of damage, as they were devastated by landslides during the tremors. Near the Piney River (just west of Nashville, Tennessee), twenty acres of land suddenly sank so low that the treetops were level with the surrounding earth.[15] As far away as Nashville, chimneys toppled from homes. In Henderson County, Kentucky, and along the Ohio River southwest of Louisville, the tremors shook down almost every brick and stone chimney in the region. At Louisville, several chimneys and gables were also toppled.

At Kaskaskia, Illinois, the earth rolled and waved violently. The church steeple bent, and the bell clanged loudly. Fissures cut through the streets, and cattle ran through town in fear. Brick chimneys toppled, and brick houses were split. Near New Athens, on the Kaskaskia River, fissures threw up white sand and water, and the chimneys of nearby homes fell. Further southeast in Illinois, in Shawneetown, chimneys also toppled from houses, and residents ran from their homes in terror. Near what is now Ridgway, a large fissure, at least two miles long and too deep for witnesses to see the bottom, opened up near a creek. The ground nearby sank two feet, and piles of white sand were vomited up in a number of places. At Jeffersonville, Indiana, the shocks set furniture in motion, and at Vincennes, the shocks were so severe that residents were concerned about whether their homes would withstand them.

At distant Natchez, Mississippi, several clocks stopped as articles fell from shelves, and as far away as Vicksburg, riverbanks caved into the waters and a number of islands were very deeply fissured and pockmarked with large sinkholes.

But accounts from regions even more distant from New Madrid (250 miles away and farther) serve to illustrate more clearly the incredible power of the December 16 shock.

Residents of Detroit felt the tremors, and on a lake thirty miles northwest of the city, Indians inhabiting an island reported that the lake "appeared to tremble, and boil like a great pot over a hot fire" and that "a vast number of large tortoises rose to the surface, and swam rapidly to the shore."[16] At Cincinnati, many residents reported a rushing or rumbling noise prior to the first tremor. (Most residents, however, were awakened by the agitation of furniture and falling chimneys.) Residents at Baltimore, Philadelphia, New York, and Newark claimed to have felt several small concussions.

At Richmond, Virginia, some of the shocks were so strong that a few residents believed their homes were being invaded by thieves, and in one mansion, bells began clanging in both the upper and lower rooms. Shocks were also felt in Norfolk and Portsmouth, where some clocks stopped, doors rattled, and articles from the ceilings of the homes swung to and fro. In Washington, D.C., furniture, doors, and windows rattled during the first tremor, which awoke many of the city's residents. As in Richmond, some residents thought they had been invaded by burglars and made room-by-room searches for the imaginary intruders. Other residents experienced dizziness and grabbed onto furniture for balance.[17]

At Raleigh, North Carolina, a number of legislators working late into the night felt the shocks at the statehouse and became alarmed by the shaking of the building, while at Georgetown, South Carolina, the early morning shocks on the sixteenth were so strong that the parade ground at the fort sank one to two inches.[18] At Columbia, the college building rocked on its foundation, and plaster fell, frightening the sleeping students enough for them to leave the building without their clothes. The air seemed to be impregnated with a vapor. At Charleston, church bells clanged, and the concussions stopped a number of clocks. Many residents experienced nausea from the shaking.

No Relief in Sight

Between December 16 and the end of the year, shocks continued throughout the New Madrid area unabated, although none was nearly as strong as the first. The ground was in a continual agitation, waving ceaselessly shock after shock. The larger tremors continued to fell trees. At one point, according to George Rogers Clark, who at the time was living on the Caney Fork River at the foot of the Cumberland Mountains in eastern Tennessee, great blocks of sandstone crashed down the mountainsides into the river.[19] Residents throughout the Midwest continued to witness flashes of light during the tremors, and the night skies seemed to be tinged with a reddish color. Fissures continued to rip and tear the land apart, usually accompanied by the escape of sulphur-tinged gas, spouts of water, and large quantities of sand, coal, and other previously buried matter. Some fissures slammed shut immediately. Others closed unexpectedly hours later. Many remained gaping open for years until debris and rainfall eventually filled them. Although some were only a few dozen feet long and a few feet wide and deep, others, twenty to thirty feet wide and of an indeterminable depth, extended for four or five miles. Sand blows, circular volcanolike formations, some of which were thirty feet in diameter, continued to vomit forth huge quantities of sand and coal.

By the end of December, the village of Little Prairie, already flooded by the relentless waters of the Mississippi, completely disappeared. In and around New Madrid, it was not uncommon for settlers, livestock, and even houses to fall into the deep fissures as they opened. Humans were eventually able to extricate themselves, but horses and cattle were usually left to die. Soon, the ingenious settlers, realizing that the fissures almost without exception ripped the land open from the southwest to the northeast, felled still-standing trees so they lay southeast to northwest on the ground, thus running perpendicular to the direction the fissures and chasms had been running. At the first sign of a new tremor, the people would run to the logs and lie on top of them, hoping that if any new fissures developed underneath, the logs would prevent them from falling into the chasms. The inventions worked, since more than once fissures did open up under the logs, and the people clinging to them were saved from falling in.[20]

The constant agitation of the ground and the relentless activity that assaulted the people's senses soon took their toll. Many people felt regular nausea and dizziness. Some took the opportunity to renew their faith. Catholics and Protestants of all nationalities knelt and prayed together throughout the ordeal, assured that the end of the world was at hand. These "earthquake Christians" experienced Christmas as they never had before.

By the end of 1811, most of the residents of New Madrid who had not already fled the area completely abandoned what was left of their homes and settlement, migrated a few miles westward to higher ground, and lived out the cold winter Indian-style in makeshift bark shelters.

Slight tremors occurred continuously into the new year, until about 8:00 A.M. on January 23, 1812, when a second violent shock, almost as powerful as the one of December 16, 1811, rumbled through the Mississippi Valley. (Many residents who had initially opted to remain in the area decided, after this second wave of devastation, to move, many of them down the river to Natchez and New Orleans.) The shock was felt with a force equal to the first, even as far away as Jamaica and Fort William Henry, New York.

Then, as if enough devastation had not been wrought, a third, and most say the most powerful and damaging, earthquake struck in the early morning hours of February 7, 1812. More than one resident later referred to it as the "hard shock" and reported that it was even stronger than the two previous major tremors. Most of the accounts of this third shock come from voyagers on the river, many of them new to the area. They reported that the river rose fifteen feet in an instant, and that "dangerous falls" and whirlpools were created.[21] The water's surface looked as though it were boiling, and huge jets of water shot skyward, taking with them mud, branches, logs, and other river-bottom debris. At New Madrid, the surging water overflowed its banks once more and completely inundated the hapless village, which at the time was in the midst of having its few remaining homes splintered to final destruction by the force of the quake below. Again, murky darkness covered the area, blocking out even morning sunlight.

Aftermath

Records indicate that the New Madrid earthquakes were felt as far away as Canada to the north, Boston to the northeast, the Atlantic Ocean to the east, Georgia to the southeast, the Gulf of Mexico to the south, and at least as far as Kansas and Nebraska (but probably farther) to the west. Seismologists estimate the shocks were felt *strongly* over an area of approximately 50,000 square miles; *moderately* over an area of between 965,000 and 1 million square miles; and at least *lightly* over an area of approximately 2 million square miles.[22] The San Francisco earth-

quake of 1906, by comparison, was felt moderately over only 150,000 square kilometers (approximately 60,000 square miles). There is a logical explanation for the extent of the area covered by the shocks. In simple terms, the earth's outer crust (the upper fifteen miles) west of the Rocky Mountains tends to soak up, and therefore diminish and dissipate, earthquake energy more efficiently than does the crust east of the Rocky Mountains. The crust east of the Rockies, in other words, is a much more efficient transmitter of the damaging wave energy of earthquakes. Subsequently, when a quake of a certain intensity strikes west of the Rockies, the earth quickly and efficiently *absorbs* the energy. When one of similar intensity strikes east of the Rockies, the earth instead quickly and efficiently *transmits* the energy great distances.

With the New Madrid earthquakes of 1811–1812, the area of greatest damage occurred, not unexpectedly, in and around the Mississippi Valley—southeastern Missouri, northeastern Arkansas, western Kentucky, and western Tennessee—an area of approximately 30,000 square miles, with the most devastating natural destruction occurring in northeastern Arkansas. Forests were particularly hard hit. Most of the trees for hundreds of square miles were either snapped, split, or uprooted. Approximately 150,000 acres of trees were specifically the victims of land that sank and was subsequently flooded, or of landslides, which were common along riverbanks and bluffs along the "bottoms" (lowlands on either side of the Mississippi River).[23]

The ground near the center of destruction was pockmarked with a number of different features, the most common of which was "sand blows," circular or elliptical mounds of sand that had been jettisoned from the ground. Most blows were eight to fifteen feet in diameter, but some reached a hundred feet. "Sand sloughs," another common feature, were shallow troughs separated by ridges of sand blown up through fissures. "Sand scatters," a third feature, were combinations of blows and sloughs. The ejection of sand occurred in two ways: quick ejection, when fissures and craters opened and spouted the sand skyward, and slow ejection, when some of the fissures closed again and extruded the sand. After the quakes, the ground between New Madrid and Little Prairie that had not sunk and become part of the river was covered with two to three feet of sand.[24] Probably the most pronounced ground destruction, however, occurred as the result of the fissures themselves, which extended throughout the area, as far as ninety miles up the Mississippi and Ohio rivers from their juncture at Cairo. The murky darkness and overpowering odors prevalent during the violent shocks were undoubtedly the results of dust agitated by the violence of the ground motion and the plethora of material—gaseous vapors, muddy groundwater, coal, sand, dirt, and organic matter—vomited up by the ground as the fissures ripped open. Some of the fissures were accurately measured at twenty feet deep in 1912,[25] a hundred years after the cataclysms, indicating that they were even deeper when they first formed, before the rain and other elements slowly began to fill them in again.

River channels in the area were decisively and permanently changed. The Mississippi River, as a result of caving banks and of uplifting and sinking (discussed below) ceased to flow along its old channel in places and instead found new routes to follow, which it still does, with few exceptions, today. The Mississippi's current course to the Gulf of Mexico, in other words, is one of the most lasting monuments to the power of the shocks. A number of other rivers, including the St. Francis River, which flows through Missouri and Arkansas, had their courses changed by the tremors. The St. Francis, in fact, used to be a well-traveled river, on which large keel boats could pass upstream from the Mississippi and portage over to New Madrid. After the quakes, it lost much of its banks, and therefore depth, and became a series of shallow swamps and bayous in places, making boat travel almost impossible.

Some of the land in the Mississippi Valley was uplifted as much as fifteen or twenty feet, in some instances emptying lakes, swamps, and rivers of their water and turning the land into dry hills or plateaus. As if in direct opposition to the uplifting, large tracts of land in northeastern Arkansas, southeastern Missouri, and even across the river in western Kentucky and western Tennessee sank, partially as a result of the settling of large alluvium deposits during the tremors, partially as a result of a reaction to the uplifting in other parts, and partially as a result of the cavities left by the ejection of sand and other organic material from the thousands of fissures throughout the area. These "sunk lands," already on the floodplains of the Mississippi River at the time, were in many places covered by water from rivers, lakes, and swamps as they sank even lower during the tremors. Some sections of land sank twenty feet or more, and the tops of the trees on that land ended up on a level with the surrounding land that had not sunk. The sinking of land, while common to the New Madrid and Little Prairie area, was not unknown in other areas. Across the Ohio River from Paducah, Kentucky, near what is now Metropolis, Illinois, a one-hundred-foot-diameter section of land is said to have sunk deep enough to leave the tops of the trees *below* the surface of the surrounding land.[26]

There is little doubt that serious environmentalists cringe in pain and horror when presented with a description of the almost total devastation of the land and the life it sustained. Recreationalists, however, at least in part hail the cataclysm. A number of beautiful lakes were created in the area as a result of the succession of uplifting and sinking. Some were created as riverbeds were uplifted and forced in different directions, leaving the old beds to form lakes. Others were created simply by depressions that filled with water from nearby streams. These latter include Lake Eulalie and Flag Lake in Missouri; Big Lake, Crooked Lake, Golden Lake, Lake St. Francis, Little Black Fish Lake, Log Lake, Sunk Land Lake, and Tyronza Lake in Arkansas; and Obion Lake and Reelfoot Lake in Tennessee. The best-known, and largest, of these is Reelfoot Lake, which, depending on how and when it is measured, is fifteen to twenty miles long, two to five miles wide, and ten to thirty feet deep. The lake was formed when a sudden uplifting of the ground dammed the water near the mouth of Reelfoot Creek, and the land to the north of the dam sank to form the new lake basin.

It is impossible to determine the precise locations of the three major shocks of 1811–1812, but seismologists, by studying accounts and descriptions of destruction, have suggested that the first major shock (December 16, 1811) was centered about sixty-five miles southwest of New Madrid, Missouri, in northeastern Arkansas. The second shock (January 23, 1812) was probably centered a few miles northwest of Little Prairie (now Caruthersville, Missouri). The third (February 7, 1812) was probably just a few miles west of New Madrid.

Chapter 3 will discuss more fully what the terms *intensity* and *magnitude* mean in relation to earthquakes, but some relative information concerning the power of the three great shocks can be given here. Earthquakes are measured on an intensity scale that ranges from one to twelve (more specifically from I to XII in roman numerals), with I being the weakest, XII being the most powerful. In terms of magnitude, while there are no upper or lower limits on the scale, over 99 percent of all quakes that occur annually measure between 3.0 and 5.0, and no measured earthquake has ever exceeded 8.9. The intensities and magnitudes represented in table 1 place the trio of New Madrid shocks among the most powerful ever occurring. It is estimated, in fact, that each shock released energy equivalent to about twelve thousand times that released by a nuclear explosion.[27]

There are no accurate accounts of the number of people killed or injured by the New Madrid earthquakes. It does seem, though, that more were killed on the Mississippi River by drowning than were killed on land. An educated guess might be a dozen or two killed on land (including Indians to the west) and four or five dozen on the river. The small number of deaths in the face of

Table 1. Comparison of Magnitudes and Intensities of New Madrid Earthquakes of 1811–1812 to Recorded Maximums

Earthquake	Richter Magnitude	Mercalli Intensity
Recorded maximums	8.9	XII
Dec. 16, 1811 (est.)	8.6	X–XI
Jan. 23, 1812 (est.)	8.4	IX–XI
Feb. 7, 1812 (est.)	8.7	X–XII

the massive destruction can be attributed to the fact that the land was sparsely settled, and the homes were small and sturdily constructed of logs.

A great many settlers moved from the area during the sequence of quakes and shortly afterward, fearing that more earthquakes would wreak havoc. Many settlements were completely abandoned. Others were left with only a handful of families, who were, as an expression of the times goes, "made of sturdier stuff." (An economic and sociological study, if it could ever be mounted, might indicate that the population of the region today would be many times what it actually is if the vast exodus had not taken place nearly two hundred years ago as a result of the winter of destruction.) Many of those who did remain continued to experience illness related to the tremors. Nausea, vomiting, and insomnia were common, and chronic fatigue resulted from lack of sleep, as residents lay awake nights fearing another catastrophic tremor. Others experienced giddiness; still others chronic depression.

The continued physical and psychological illnesses, combined with the initial trauma of the first violent tremor and the fear that the world might really be ending, was enough to create a new wave of religious fervor in the region. Wayward settlers dusted off their Bibles and searched feverishly for passages related to the destruction of the world; and traveling preachers, hearing of the need for their services, flocked to the Mississippi Valley in 1812. Church attendance doubled in some communities. Within a few years, however, the shocks had subsided with no indication of return, and church attendance and interest in religion concomitantly declined. Frustrated preachers, trying to maintain the fervor from days past, began referring to these backsliders as "earthquake Christians" and cajoled them to return to the flock. Their efforts, for the most part, were fruitless. The danger, it seemed, was over.

Mild shocks continued to visit the area for almost a decade after the major ones had ceased, and the residents who remained became used to them, often not even interrupting their activities to wait the tremor out.

What Really Happened?

History is usually only as good as contemporary accounts and, to some extent, the on-location observations of later researchers. Accounts of the effects of the 1811–1812 New Madrid earthquakes are available from three sources: the writings of eyewitnesses, the writings of people who communicated with eyewitnesses, and the reports of others (most commonly scientists) who visited the area years later and observed the aftermath of the destruction. Each source, in its own way, is subject to error.

Eyewitnesses, for example, were capable of only being in one place at one time, so their observations were limited. To help the cataclysm make sense to them, it is not impossible that they made assumptions about the effects of the devastation elsewhere and reported those as facts. Since few, if any, of the people living in the area at the time had experienced earthquakes before, the activity must have seemed more like the end of the world than a natural event, so accounts have undoubtedly been marred by concerns with more than just the events themselves. The fear that gripped the witnesses to the destruction must have played a role in their accounts, for instance. This phenomenon is best stated by a sailor, who, being queried about the shocks years later, stated, "When a man expects nothing but instant death it is hard for him to think or notice anything but his danger."[28]

Many reports are, of course, from people who did not actually witness certain events, but heard about them from others. Much can obviously be lost in this kind of translation. The difference between what actually happens and what is reported to a second or third person can be great. Similarly, even eyewitnesses can exaggerate situations to make them seem more substantial than they really are. Rumors and exaggerations are two of history's most common enemies.

Time has a way of leading to selective retention and even gross error. Most comprehensive descriptions of the New Madrid earthquakes, for example, include the account of the naturalist John James Audubon, who was supposed to have been living in western Kentucky at the time. Audubon's account, however, written years later, is suspect. He states, first of all, that the first shock occurred in November of 1812, when in fact in occurred in December of 1811. (There are no accounts or records anywhere of shocks in November 1812.) He then goes on to state that it occurred in the afternoon while he was riding his horse. None of the three shocks occurred in the afternoon; all took place in the early morning hours.[29]

It may well be, however, more important ultimately to examine the causes for such catastrophes than to dwell on reports of their effects. The violence that beset the Midwest in the winter of 1811–1812 was only part of a series of earthquakes and volcanoes that occurred during that time period. As mentioned in the first part of this chapter, there were a number of violent shocks and eruptions in early 1811. There were more in 1812 and 1813 throughout North, Central, and South America. On March 26, 1812, major shocks destroyed Caracas and La Guaira in Venezuela and killed between ten thousand and twenty thousand people. On December 8, 1812, San Juan Capistrano, California, was destroyed by a quake; and four days later, Purisima, California, was also hit. Whether all of these quakes and volcanoes were related or were triggered by a separate mechanism (that is, whether there is a connection between earthquakes and volcanoes in different parts of the world) will be discussed in chapters 3 and 4, but first an overview of Mississippi Valley earthquake activity during the years both preceding and following the quakes of 1811–1812 is in order.

2. Centuries of Mississippi Valley Earthquakes

Then and Now

While geological evidence tells us earthquakes occurred in the Mississippi and Ohio River valleys for eons before settlers moved in and began recording the activity, the first recorded shock took place on Christmas Day, 1699, on the Mississippi River near the present-day site of Memphis. A second shock is said to have occurred along the river a few days later, January 4, 1700.[1]

Earthquake activity, however, never rests in the Mississippi Valley, and earthquakes continue to visit the region on a regular basis, even today. Most tremors are so weak, though, that they are rarely felt. Their presence is recorded only by sensitive seismological instruments. In 1982, for instance, seismograms recorded approximately twenty thousand small earthquakes in a two- or three-square-mile area of northern Arkansas. Fifteen hundred were recorded in one day alone.[2] In February 1984, small shocks were felt in southern Illinois and western Kentucky, surrounded by a series of even smaller foreshocks and aftershocks. Most of this activity is what seismologists refer to as "swarm earthquakes," small regional clusters of very mild tremors occurring over a period of time, usually a week or so.

The first really strong shock to hit the area since the New Madrid earthquakes of 1811–1812 occurred in southwestern Illinois east of St. Louis on June 9, 1838 (see table 2 and map 1). Its intensity is estimated to have been between VII and VIII, and

Table 2. The Most Powerful Mississippi and Ohio Valleys
Earthquakes Since 1811–1812

Date	Richter Magnitude	Mercalli Intensity	Affected Area (Sq. Miles)
June 9, 1838	5.7	VII–VIII	200,000
Jan. 4, 1843	6.0	VIII	400,000
Oct. 8, 1857	5.3	VII	7,500
Aug. 17, 1865	5.3	VII	24,000
Nov. 18, 1878	4.9	VI	150,000
Apr. 12, 1883	5.0	VI–VII	(small)
Oct. 31, 1895	6.2	IX	1,000,000
Apr. 29, 1899	5.0	VI–VII	40,000
Nov. 4, 1903	5.3	VII	70,000
Aug. 21, 1905	5.0	VI–VII	40,000
Sept. 27, 1909	5.3	VII	30,000
Apr. 9, 1917	5.0	VI	200,000
Nov. 26, 1922	5.0	VI–VII	50,000
Oct. 28, 1923	5.3	VII	40,000
Apr. 26, 1925	5.0	VI–VII	100,000
May 7, 1927	5.3	VII	130,000
Aug. 19, 1934	5.0	VI–VII	28,000
Nov. 23, 1939	4.9	V	150,000
Mar. 3, 1963	4.7	VI	100,000
Aug. 14, 1965	3.8	VII	300
Oct. 20, 1965	4.9	VI	160,000
Nov. 9, 1968	5.5	VII	580,000
Apr. 3, 1974	4.7	VI	243,000
Mar. 25, 1976	4.9	VI	108,000

NOTE: Earthquakes listed are those with magnitudes of at least 5.0, intensities of at least VII, or ranges of at least 100,000 square miles.

SOURCES: Otto W. Nuttli, "Magnitude-Recurrence Relation for Central Mississippi Valley Earthquakes," Bulletin of the Seismological Society of America, 64 (Aug. 1974): 1191–96; Jerry L. Coffman, Carl A. von Hake, and Carl W. Stover, Earthquake History of the United States, rev. ed., prepared for the U.S. Department of Commerce (Washington: GPO, 1982), 38–43, 11a–12a; and Otto W. Nuttli, "Damaging Earthquakes of the Central Mississippi Valley," in Investigations of the New Madrid, Missouri, Earthquake Region, Geological Survey Professional Paper no. 1236, ed. F. A. McKeown and L. C. Pakiser (Washington: GPO 1982), 16–18.

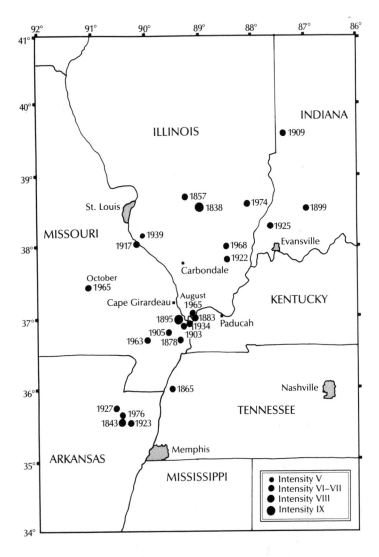

Map 1. Approximate Epicenters of the Most Powerful Mississippi and Ohio Valleys Earthquakes Since 1811–1812

it was felt over an area of approximately 200,000 square miles, including parts of Kentucky and Indiana. In St. Louis, residents in upper stories of buildings felt the shock noticeably.[3]

Other "moderate" earthquakes have followed. The shock of September 27, 1909, intensity VII, was centered near the Wabash River along the Indiana-Illinois border and was felt over approximately 30,000 square miles. In Terre Haute, Indiana, two chimneys toppled and some plaster cracked. The tremor was felt in southern Indiana, southern Illinois, western Kentucky, northern Arkansas, and parts of Iowa and Missouri, as far west as Kansas City. Centered near Ste. Genevieve and St. Mary's, Missouri, the earthquake of April 9, 1917, intensity VI, was felt over approximately 200,000 square miles, including Missouri, Illinois, Iowa, Wisconsin, Indiana, Kentucky, Tennessee, Mississippi, Arkansas, and Kansas. Throughout the area, chimneys toppled, plaster cracked, buildings shook, windows rattled, and furniture moved. The August 19, 1934, shock was centered in southeastern Missouri near Cairo, Illinois. It had an intensity of between VI and VII and was felt over approximately 28,000 square miles, including southeastern Missouri, southern Illinois, western Kentucky and Tennessee, and northeastern Missouri. At Charleston, Missouri; Cairo, Illinois; Mound City, Illinois; and Wickliffe, Kentucky, windows were broken, chimneys toppled, and articles were thrown from shelves.

Although centered in two different parts of the region, two 1965 earthquakes were significant. That of August 14, 1965, was centered near the southern tip of Illinois and had an intensity of VII; it was only felt, however, for 300 square miles. At Tamms, Illinois, chimneys fell, walls cracked, groceries fell from shelves, and water supplies were muddied. The tremor of October 20, 1965, was centered in eastern Missouri, southwest of St. Louis. This quake, with an intensity of VI, was felt over approximately 160,000 square miles including parts of Arkansas, Illinois, Iowa, Kansas, Kentucky, Tennessee, Oklahoma, and Nebraska. Near the epicenter, the ground swayed, walls cracked, dishes and windows rattled, bricks fell, and furniture rocked.

Danville Public Library
Danville, Illinois

Three Major Shocks

The earthquake of 8:45 P.M., January 4, 1843, centered in north-eastern Arkansas just west of Memphis, Tennessee, is the fifth most powerful earthquake known in the Mississippi Valley. Only the 1811–1812 series and the 1895 shock (discussed below) were stronger. The 1843 earthquake's intensity is estimated to have been VIII, and it was felt over 400,000 square miles (most of the Midwest). Besides Arkansas, the tremor was felt in Missouri, Illinois, Kentucky, Tennessee, Indiana, Ohio, West Virginia, North Carolina, South Carolina, Georgia, Alabama, Mississippi, Louisiana, and possibly Michigan, Texas, Oklahoma, and Nebraska. Chimneys fell and brick walls cracked throughout the region. The tremor was so strong in Louisville, Kentucky, that residents reported having difficulty standing. Land sank near New Madrid, Missouri, and along the St. Francis River in northern Arkansas, a new lake formed when part of the riverbed sank. In and around Memphis, building foundations and walls cracked, windows broke, chimneys fell, and buildings shook. The tremor was preceded and accompanied by rattling and rumbling sounds, and hundreds of the city's residents ran into the streets in panic. Outside an auction house that was being unceremoniously evacuated as a result of the shaking, a number of people were trampled and crushed by the frightened crowd.

The strongest earthquake since the 1811–1812 New Madrid trio occurred a little after 5:00 A.M., October 31, 1895, near the town of Charleston, Missouri. It had an intensity of at least VIII and was felt over a million square miles and twenty-eight states: Arkansas, Missouri, Iowa, Minnesota, South Dakota, Nebraska, Kansas, Oklahoma, Texas, Louisiana, Mississippi, Alabama, Georgia, Florida, South Carolina, North Carolina, Tennessee, Kentucky, Illinois, Wisconsin, Michigan, Indiana, Ohio, West Virginia, Virginia, Pennsylvania, Maryland, and New York. At Charleston, Missouri, sand blows were reported, spouting sand and water to a height of three feet; fissures ejected sand, coal, and water, and then filled with water; and four acres of ground sank, creating a new lake near the city. In St. Louis, most of the city's residents were awakened by the shock, many of them being thrown from their tossing beds, as houses and buildings swayed, creaked, and cracked. Furniture moved about, chimneys fell, as did other bricks from walls, and a number of ceilings and foundations were cracked.

The most recent major shock—the earthquake of November 9, 1968, centered in southeastern Illinois—remains this century's most powerful Mississippi and Ohio valleys quake. With an intensity of VII, it was felt over almost 600,000 square miles, and reports of its effects came in from twenty-four states: Illinois, Indiana, Kentucky, Tennessee, Missouri, Iowa, Arkansas, Mississippi, Alabama, Georgia, Florida, South Carolina, North Carolina, Virginia, West Virginia, Pennsylvania, Ohio, Michigan, Wisconsin, Minnesota, Nebraska, Kansas, Oklahoma, and Massachusetts. There were even reports of shaking at Toronto, Canada. In terms of ground motion, residents close to the center of the earthquake reported seeing the ground wave, roll, and ripple slightly. Waves formed on many area lakes, and observers reported fish jumping out of the water. In places, sidewalks, pavement, and the ground cracked slightly. Waterlines broke, and at one town a water tank split and sent a jet of water fifty feet into a nearby parking lot. Dozens of tombstones in local cemeteries either toppled or rotated from their mounts. Throughout the region, power and telephone lines and poles swayed. Although automobile drivers felt the tremor, they were not as severely affected as many drivers of large trucks, who had to pull off the roads until the shock subsided. At St. Louis, police reported automobiles skipping several inches in a parking lot.

Houses throughout the region experienced slight to severe damage. Chimneys toppled from hundreds of homes, windows broke, plaster cracked and crumbled, foundations cracked, bricks popped loose from walls, antennas fell, furniture shifted, objects fell from shelves, and chandeliers swung. Retailers throughout the area reported slight to moderate damage. Merchandise tumbled from shelves in a number of grocery stores, up to one third of the merchandise in some stores. Buildings throughout the region and across the Midwest and eastern states felt the effects of the tremor. Plaster and brick cracked and fell, windows broke, and heavy furniture slid across floors. Buildings also swayed in Memphis, Omaha, Chicago, Milwaukee, La Crosse (Wisconsin), and Grand Rapids. People in tall buildings in Toronto and Boston also reported feeling the tremors. Throughout the Midwest, reaction was mixed. Many people were extremely frightened and ran from their homes. Telephone switchboards, police and fire stations, and media offices throughout the cities in the region were swamped with calls of concern. Others noticed the tremors and were perplexed. Still others realized what was happening and thought little about it. A few people found it difficult to walk during the tremor. Injuries, miraculously, were few. Specific reports indicate that only four people throughout the Midwest were injured as a result of the earthquake, although there might well have been more.

There is at least the possibility that the next New Madrid earthquake will become this century's most powerful shock. Every effort must be exerted by the heretofore ill-prepared region to not allow damages and injuries (including fatalities) to approach or exceed those of the great earthquakes of 1811–1812.

PART TWO: **THE PRESENT**

3. Earthquake Theory

The Trembling Planet

For what consolation it may offer, Planet Earth is not the only heavenly body in the solar system that shakes due to internal activity. Scientists have found that our closest neighbor in the heavens, the moon, also quakes on occasion. Although certainly not as powerful as earthquakes (their average magnitude is a paltry 2.0 on the Richter scale, hardly powerful enough to be felt by humans), these moonquakes, as they logically enough are called, occur with surprising frequency. Seismic stations deployed on the moon during the 1969 Apollo mission recorded over 1,000 moonquakes in their first eight years, an average of over 125 per year.

Unless and until the moon is colonized or industrialized, however, people on earth will have little interest in moonquakes. There is enough quake activity on earth to keep us busy—and concerned—here. Powerful tremors can cause hundreds of thousands of deaths, depending on when and where they strike. The most devastating shock in terms of numbers of people killed is thought to have occurred in China in 1556, when an estimated 830,000 peasants were killed as an earthquake crumbled the bluffs above the caves they were living in and crushed and buried them (see table 3). As recently as July 28, 1976, an estimated 240,000 Chinese were killed when an earthquake measuring 8.2 on the Richter scale caught the residents of heavily populated Tangshan near the capital of Beijing by surprise. (Officials had predicted another powerful earthquake just a year earlier and safely evac-

Table 3. The World's Most Powerful Earthquakes

Location	Date	Richter Magnitude	Mercalli Intensity	Number Killed (Estimated)
China	1556	—	—	830,000
India	1737	—	—	300,000
U.S. (New Madrid)	1811–1812	8.4–8.7	X–XII	Few
Japan	1857	—	—	107,000
U.S. (San Francisco)	1906	8.3	XI	1,500
Colombia/Ecuador	1906	8.9	XII	1,000
Italy	1908	—	—	160,000
China	1920	8.6	XI–XII	150,000
Japan	1923	8.3	XI	143,000
China	1927	8.3	XI	200,000
Japan	1933	8.9	XII	2,900
India	1950	8.7	XII	1,530
Chile	1960	8.9	XII	5,700
U.S. (Alaska)	1964	8.4–8.5	XI	131
China	1976	8.2	XI	240,000

NOTE: Earthquakes listed are those with magnitudes of 8.0 or more or related fatalities of 100,000 people or more.

SOURCES: James Cornell, *The Great International Disaster Book* (New York: Pocket Books, 1979), 131–52; Encyclopaedia Britannica Editors, *Disaster! When Nature Strikes Back* (New York: Bantam Books, 1978), 19–41; "Chinese Earthquake: Latest in a Chain That Raises Worldwide Concern," *U.S. News and World Report*, 9 Aug. 1976, 18–23; and Otto Nuttli, letter to author, March 16, 1987.

uated hundreds of thousands of residents from the area before it struck.) The tremor also left another 780,000 injured and 500,000 homeless. In 1737, a powerful shock in India left an estimated 300,000 dead in that country. In Japan, a shock measuring 8.3 led to the deaths of 143,000 in 1923. Initial studies report that the September 19, 1985, earthquake on Mexico's western coast (a 7.8 on the Richter scale), left 6,000 dead, 2,000 missing, 30,000 injured, 150,000 homeless, and caused approximately $4.1 billion damage. In the twentieth century, the ten most devastating earthquakes alone have left over 1.5 million people dead.[1]

In the United States, the three most powerful earthquakes since the New Madrid trio in 1811–1812 were the 1886 tremor that shook Charleston, South Carolina, the 1906 San Francisco shock, and the 1964 Alaskan temblor. The Charleston upheaval did $23 million damage (an almost unheard of amount in 1886).[2] The San Francisco cataclysm, measuring 8.3 on the Richter scale, left as many as 1,500 dead, 80 percent of them as a result of the ensuing fires that gutted the city. (Early estimates suggested approximately 500 dead. More recent government figures estimated 700. However, an archivist in the city studying the data claims to have the names of 1,500 victims, noting that earlier "official" figures ignored the deaths of hundreds of poor and indigent newcomers to the city.)[3] The 1964 disaster on Alaska's coast, measuring 8.4 to 8.5, left just 131 dead, 122 of them victims of the resulting *tsunami* (a huge wave, often mistakenly called a tidal wave, triggered by the upheaval of the ocean bed at or near an earthquake's epicenter).

Frequencies of earthquakes and totals of deaths and damage caused by them over the years are equally as frightening as individual events. Possibly as many as thirteen million people have died in earthquakes in the past four thousand years. Some researchers suggest seventy-five million have died as a result of earthquakes since the beginning of humankind.[4]

On an annual basis, approximately one million earthquakes occur around the world—an average of one tremor every thirty-two seconds (see table 4). Only about 2,000 to 6,000 of these,

Table 4. Earthquake Magnitude, Intensity, and Annual Incidence

Richter Magnitude	Maximum Mercalli Intensity	Expected Annual Incidence
Below 3.0	I	1,000,000
3.0–3.9	II–III	49,000
4.0–4.9	IV–V	6,200
5.0–5.9	VI–VII	800
6.0–6.9	VII–VIII	120
7.0–7.9	IX–X	18
8.0–8.9	XI–XII	Fewer than 1

SOURCE: "Magnitude and Intensity: Measures of Earthquake Size and Severity," *Earthquake Information Bulletin,* Sept.–Oct. 1974, 24.

however, are powerful enough to be felt, and only 800 strong enough (magnitude 5.0 to 5.9) to cause some damage to humanmade structures. About 120 of these are devastating enough (magnitude 6.0 to 6.9) to cause major damage to structures, and 18 or so (magnitude 7.0 to 7.9) are powerful enough to destroy a major city (if they were to strike one, which, fortunately, not many do). Approximately every three years, one of the most powerful earthquakes (magnitude 8.0 and above) shatters a region. Earthquakes leave an average of between ten thousand and fifteen thousand dead each year and inflict nearly $7 billion of damage worldwide.[5] In the United States alone, the ten most destructive shocks since the turn of the century have killed around two thousand people and caused close to $5 billion in damage.[6]

The recent increase in media attention seems to suggest that earthquakes are becoming more commonplace. Scientists report, however, that the numbers, intensities, and frequencies of earthquakes around the world remain about the same. What is changing is the population. As the world's population grows and continues to build on or near the earth's known earthquake regions, the amount of damage from earthquakes grows also. In addition, many of the more recent temblors (those since 1980) have occurred in or near large cities and have thus received greater media attention. Actual deaths from earthquakes did decrease for a three-year period, though. In 1982, 3,338 people were killed; in 1983, 2,322; and in 1984, 770, the lowest death toll since the 1940s.[7] However, there was a significant increase in deaths in 1985, primarily as a result of the Mexico earthquake.

For the most part, U.S. citizens do not seem to be very concerned about the threat of earthquake damage, even though most of the nation is at some kind of risk. Approximately 70 million U.S. citizens live in areas where the risk is deemed *serious,* and another 120 million live in areas where the risk is deemed *moderate.*

Earthquake Damage

Much of the damage earthquakes cause results from the direct hazards they present (see table 5). One of these hazards is *fault*

Table 5. Types of Earthquake Damage

Direct	Indirect
Fault displacement	Fire
Tectonic uplift and subsidence	Ground failure
Seismic waves	Fissuring
Body waves	Liquefaction
P waves	Slope failure (landslides)
S waves	Compaction
Surface waves (L waves)	Quick condition failure
Rayleigh waves	Water activity
Love waves	Seismic seiches
	Tsunamis

displacement, which is the pulling apart, moving together, or diagonal shifting of the land on either side of a fault. Structures that straddle the fault, in other words, will be rent apart, crushed together, or ripped sideways as the fault causes direct ground motion. A second direct hazard is *tectonic uplift and subsidence,* during which the bedrock, and the soil atop it, is actually raised or lowered during an earthquake. This sudden shifting can cause considerable damage to structures in the area.

The most serious and widespread damage that results from direct earthquake hazards, however, results from ground shaking. Ground shaking consists of a number of different types of *seismic waves* that pulse through the ground and across the surface of the earth. To understand the destructive nature of these waves, it is first necessary to understand specifically where earthquakes occur. The center of an earthquake, which is always below the surface of the ground, is called the *hypocenter.* The point of land directly above the hypocenter, on the earth's surface, is called the *epicenter.* Earthquakes can occur at various depths. Shallow earthquakes are defined as those that occur between 0 and 43 miles below the surface of the ground, intermediate earthquakes as those that occur between 43 and 186 miles, and deep earthquakes as those that occur between 186 miles and 434 miles below the surface of the ground.

When an earthquake occurs, two types of waves (called *body waves*) emanate from the hypocenter and work their ways toward the surface. On the one hand, P (primary or pressure) *waves* act like sound waves and pulse through the ground, compressing and dilating the rock and pulling and pushing the earth, eventually causing up-and-down motions as they reach the surface, much like an upended stereo speaker cone would pulsate up and down as a result of the magnetic current below it. In fact, P waves are responsible for the various sounds (such as rolling thunder and booming sounds) reported during earthquakes. On the other hand, S (secondary or shear) *waves* advance about half as quickly as the P waves and snake through the ground like whips, shaking the rock and the earth sideways. When S waves reach the surface, they cause a back-and-forth motion on the ground. Together, P and S waves can result in violent ground shaking so severe that trees can snap, buildings can crumble, and people can suffer broken limbs while trying to walk.

There are also two types of waves, called *surface waves* or *L* (long) *waves*, that originate on the surface of the ground and travel much slower than body waves. They also travel much farther than body waves, however, and generally cause the most damage. *Rayleigh waves* churn across the surface of the ground vertically like ocean waves. *Love waves* vibrate at right angles to their direction of travel, whipping back and forth horizontally. Together, the two types of waves create a corkscrew motion that can cause significant damage. L waves cause structures to sway and buckle (most structures are damaged because they are not designed to withstand *horizontal* ground motion) and often trigger landslides.

Earthquakes also result in a number of indirect hazards that can cause equally serious damage. Obvious among these is fire, which can start from any number of sources (such as ruptured gas lines, downed power lines, and overturned flame sources in homes), and the type of *ground failure* called *fissuring*, the opening of cracks in the ground as a result of the weakening and breaking up of claylike soil from the various waves moving through it.

A second, less obvious, type of ground failure is *liquefaction*, the transformation of sandy (noncohesive), water-saturated soil from a solid state to a liquid state not unlike quicksand. Liquefaction is a consequence of increased pore water pressure, which causes the soil to lose its stability and react like gelatin when agitated by seismic waves. During liquefaction, subsurface water can be ejected into the atmosphere in the form of such phenomena as sand blows. One consequence of liquefaction is *slope failure* (or landslides). A second is vertical *compaction* (also, confusingly, called subsidence), in which unconsolidated, liquefied soil sinks as a result of its granules becoming compacted. This phenomenon is different from tectonic subsidence (described earlier), which is the actual lowering of the bedrock as a result of earthquake activity. (Vertical compaction is the result of settling soil, in other words, not of depressed bedrock.) A third result of liquefaction is *quick condition failure*, in which structures atop the liquified soil sink into it, as they would into quicksand, *and* in which some buried structures (tanks, sewage lines, and other vessels, for instance) may rise (float) to the surface. During liquefaction, homes and other buildings can sink into the ground, or parts of them may sink, causing the structures to lean precariously and even topple over completely.

The ground is not the only place agitation occurs, of course. Bodies of water are equally susceptible to activity as a result of earthquakes. Earthquakes can, for example, cause agitation on smaller bodies of water, such as lakes, rivers, and bays. This agitation, a series of waves set in motion by the tremors, is called a *seismic seiche,* or simply a seiche. Seiche waves rarely reach more than five or six feet in height. They can occur, however, at great distances from the epicenter of an earthquake. For example, some of the most active seiche activity during the 1964 Alaska earthquake occurred in the lakes and reservoirs in the southeastern United States. In fact, waves of a foot or more in height were reported on Lake Ouachita near Hot Springs, Arkansas. Seiches are capable of causing flooding and mild to moderate damage to boats, harbors, and other structures near the shores. Tsunamis (briefly discussed earlier) are, on the other hand, seismic ocean waves (often incorrectly called tidal waves), that, precipitated by underwater earthquakes, occur on the oceans. These waves, unlike seiche waves, can reach frightening heights (forty feet or more) by the time they hit, and are capable of wiping out seacoast communities. The 1964 Anchorage, Alaska, earthquake is an example of devastating tsunami damage.

Earthquake Measurement

There are two scales commonly used to measure earthquakes, and they are often confused. One is the Modified Mercalli Intensity Scale of 1931, which measures the *degree of ground shaking* caused by an earthquake at a certain location (see appendix 1). The second is the Richter Magnitude Scale, which measures the *amplitude* of an earthquake, which in turn can be translated into the *amount of energy released* by the quake (see table 6).

The Mercalli Intensity Scale, since it measures damage at various locations hit by the tremors, is a subjective scale that measures the severity of ground motion based on the reports of observers at the various locations. In other words, an earthquake has its greatest intensity at its epicenter. Farther away from the epicenter, as intensity decreases, so does damage. The Mercalli scale, which uses roman numerals to designate its twelve levels, was developed in 1902 by the Italian priest and

Table 6. Earthquake Magnitude and Energy Release

Richter Magnitude	Approximate Energy (*In Equivalents of TNT*)
1.0	6 ozs.
2.0	13 lbs.
3.0	400 lbs.
4.0	6 tns.
5.0	200 tns.
6.0	6,270 tns.
7.0	100,000 tns.
8.0	6,270,000 tns.
9.0	200,000,000 tns.

SOURCE: William J. Petak and Arthur A. Atkisson, *Natural Hazard Risk Assessment and Public Policy: Anticipating the Unexpected* (New York: Springer-Verlag, 1982), 28.

geologist Giuseppe Mercalli and later modified in 1931 by American seismologists H. O. Wood and Frank Neumann to include descriptions of damage to more modern structures. To determine the intensity of an earthquake throughout the region it affects, the United States Geological Survey sends questionnaires to volunteer observers in that region (among them, postmasters, forest service personnel, and weather service observers), asking them to describe damage. Once all the information is gathered, maps are drawn to visually show the intensity of the earthquake throughout various points in the region. These are called isoseismal maps.

The Richter Magnitude Scale, again, measures the amplitude of an earthquake at its moment of creation. Magnitude, unlike intensity, which is measured by subjective observations, is measured directly by instruments called seismometers at seismographic recording stations throughout the world. The graphs produced by the delicate instruments are called seismograms. The scale was developed by Charles F. Richter in 1934. (Richter, often called the "father of modern seismology," died in 1985.) The Richter scale is actually open-ended. That is, it technically has no upper or lower limits. The smallest earthquakes, however, usually measure slightly above zero, and the largest ever recorded measured 8.9 on the scale—which is probably very close to the upper limit, since the planet's materials and activities are incapable of generating and holding energy much greater than that measured as 8.9.[8]

Also unlike the Mercalli scale, the Richter scale is logarithmic, meaning that each increase of one magnitude level corresponds to a *10-fold* increase in vibrational amplitude. (For example, an earthquake measuring 6.0 on the Richter scale has ground waves ten times stronger than an earthquake measuring 5.0 on the same scale. An earthquake measuring 3.2 is ten times stronger than one measuring 2.2.) In addition, each increase of one magnitude level corresponds to a *31.5-fold* increase in energy release. An earthquake measuring 1.0 can generally only be detected by a seismometer. People can often slightly detect one measuring 2.0. A 5.0 magnitude earthquake can damage structures, and those measuring 6.0 are capable of severe damage. Earthquakes measuring between 7.0 and 8.9 on the scale are truly catastrophic. (For a listing of the world's most powerful earthquakes, including their Mercalli and Richter estimates, see table 3. For a comparison of Mercalli and Richter, see table 4.)

The Turmoil Under Our Feet

In the fifth century B.C., Democritus suggested that earthquakes were caused by rainwater seeping into the ground. Aristotle later argued they were caused by pockets of gas escaping from underground cavities. Throughout the centuries, various other theories have come forward, many of them pointing to supernatural powers; only within the last twenty years have the mechanics of earthquakes really begun to be understood.

New information will certainly be modified, expanded on, or even abandoned over the years (thus, what science thinks it knows today may cause smiles among future, more knowledgeable, generations, just as we smile today as we read century-old theories), but at present it is suggested that the earth is covered by seven major, and between twelve and twenty minor, hard and rigid *plates*. These plates are approximately fifty to sixty miles thick, and the largest are millions of square miles in area. In many instances, these plates follow continental and oceanic lines. The North American Plate, for instance, extends from the western coast of North America, where it meets the Pacific Plate, to the middle of the Atlantic Ocean, where it meets the Eurasian Plate. The Eurasian Plate extends eastward, where its eastern edge meets with the western edge of the Pacific Plate. The South American Plate meets the African Plate in the middle of the Atlantic Ocean and meets the Nazca Plate (a smaller plate between the South American and the Pacific Plates) on its western edge. Other smaller plates include the Antarctic Plate, the Cocos Plate (just off Mexico's western coast), the Caribbean Plate, the Scotia Plate (just off Antarctica), the Arabic Plate, the Indian-Australian Plate, and the Philippine Sea Plate.

These plates "float" on the top, plasticlike fluid layer of the earth's mantle, which extends approximately eighteen hundred miles deep, where it meets the earth's core. The term "floating" is crucial to understanding earthquakes and the branch of science (known as plate tectonics) that studies these floating plates.[9] Plates, that is, can move, albeit slowly, over the mantle. To understand the concept more clearly, one can imagine a patio that has been built with flat rocks that are not cemented together. Over time, as the ground below the patio is subject to moisture from rain as well as to other slight movements of the earth, the rocks will gradually shift position. Some will move closer to one another and move farther away from others. Still others will begin to tilt slightly upwards or downwards, victims of the activity of the ground below them.

Scientists have found that along midoceanic plate boundaries (the edges of plates that lie underneath some of the oceans, such as the Atlantic Ocean), the mantle is constantly ejecting molten rock material up through the cracks between the plates, forcing them apart. The ocean floors, thus, are constantly being renewed and continually spread apart, forced toward the continents. As the edges of the oceanic plates reach the edges of the continental plates, they dive under these more stable plates, and the cool oceanic crust moves back toward the center of the earth, where it is melted in the mantle. This phenomenon is not unlike a system of conveyor belts or escalators moving horizontally. After the belts or steps arise from below the floor, they move slowly over its surface and then disappear again. As the belts or steps disappear below the floor and reappear at the far end of the mechanism, they melt and then form and become cool again. This process is similar to how water boiling in a pot moves toward the surface, where it is temporarily cooled, and then dives again toward the bottom to be reheated.

This movement goes far to explain and prove the theory of continental drift, first proposed in 1912, which suggests that North and South America were once part of one larger continent containing Europe, Asia, and Africa. In other words, as convection occurred in the middle of what is now the Atlantic Ocean—

which was once much narrower than it is today—the continents were slowly pushed apart, North and South America on a "conveyor belt" heading west, and Europe, Asia, and Africa on a "conveyor belt" heading east. The process has taken almost two hundred million years. A resulting phenomenon has been that the North American and South American plates are being pushed farther westward, into the Pacific, Cocos, and Nazca plates. (By using examples like conveyors and escalators, there is a risk of giving the impression that plates move rather rapidly. This is simply untrue. Plates move over the mantle at incredibly slow speeds—between one and five inches per year on average.)

When plates meet, a number of things can happen, depending on where they meet. Essentially, there are four possibilities: (1) When some plates meet, one will dive under the other (this is called *subduction* and is described above). An example of this process occurs along the western coast of Mexico and South America, where the oceanic plates dive under the continental plates and form deep ocean trenches. (2) A second possibility for activity exists along plate edges that meet underneath oceans, but away from more stable plates. Here, as explained earlier, molten mantle exudes between the plates and forces them apart in opposite directions. (3) When other plates meet, their edges buckle, move upward, and create mountains as a result of the compression. An example of this type of activity occurs where the Indian-Australian Plate meets the Eurasian Plate at the southern edge of Asia, as they continue to form the Himalaya and Caucasus mountains. (4) When still other plates meet, especially if they meet at an angle rather than head on, they will glide past one another in a grinding motion. These areas make up one type of *fault*. The most well-known example is the San Andreas Fault in California, where the eastern edge of the Pacific Plate is slowly grinding along the western edge of the North American Plate, with the former moving northwest and the latter moving southeast.

When any of this activity occurs, pressure builds up, and the result is an earthquake. In the first instance, the surface of a diving oceanic plate can become caught on the bottom of the edge of a continental plate and build up pressure as it continues to try to move. Eventually, when enough pressure is built up, the rock weakens and buckles, and an earthquake results. The pressure is then alleviated, and the plates again move above and below each other. In the second instance, earthquakes can occur below ocean floors as the molten rock being forced up between the plates puts pressure on the plates. The earthquakes relieve the pressure, and the ocean floors can continue their convey-orlike movement toward the continents. In the third instance, earthquakes can occur when plates moving directly against one another build up enough pressure to cause the rocks to buckle. When the rupture occurs, the rocks are forced upward, and the mountains can continue to build. In the last instance, the plates that are grinding past one another build up friction. As the pressure builds, an earthquake results at the pressure point. The earthquake then releases the pressure and allows the plates to continue grinding past one another.

Not surprisingly, then, 90 percent of the world's earthquakes occur along the edges of the earth's tectonic plates. As the rocks along the plate edges continue to receive the pressure of the plates, they begin to crack, swell, dilate, and bulge. Groundwater then begins to seep into the rocks to fill the microcracks. As this occurs, the rocks begin to lose their shear strength and eventually rupture. This rupturing is an earthquake.

The Midwest

Midwestern Geology

The previous discussion is helpful in explaining how over 90 percent of the earthquakes that take place around the world occur. It does little, however, to explain the other 10 percent, including midwestern earthquakes. The midwestern United States does not lie on, or anywhere near, a plate boundary. In fact, the region is right in the middle of the North American Plate, hundreds of miles from both the eastern edge (the middle of the Atlantic Ocean) and the western edge (the eastern edge of the Pacific Ocean).

The Midwest is not the only area distant from plate edges to experience earthquakes. South Carolina, Massachusetts, Newfoundland, the St. Lawrence Valley, and Montana are other areas well within the North American Plate to have experienced strong earthquakes. Such earthquakes are referred to as "intraplate" earthquakes, because they occur within the boundaries of known plates.

The Midwest, however, is the most seismically active area east of the Rocky Mountains. In fact, in some ways, it rivals the more well-known California earthquake activity. The three-month sequence of earthquakes occurring near New Madrid in 1811–1812 (including the approximately one thousand intermittent tremors and aftershocks), for instance, equalled the size and number of *all* earthquakes that occurred in southern California during the forty-year period from 1932 to 1972.[10] To this day, tremors measuring at least 1.0 on the Richter scale still occur at least three or four times a week in the Mississippi Valley.

As map 2 shows, the Midwest, especially the area around the Mississippi Valley, is laced with a series of faults (cracks or breaks in the continuity of rock formations). The genesis of these faults goes back almost five hundred million years ago, when the "proto-Atlantic" (the precursor to the Atlantic Ocean) was being formed to the east by the tearing open of the continental crust.[11] The crust in the New Madrid area, experiencing the same stress, pulled apart slightly and formed a narrow, sunken rift (called the Reelfoot Rift) that runs down what is now the course of the Mississippi River from around its juncture with the Ohio River southward to the Gulf of Mexico. The Reelfoot Rift, which is approximately twenty-five miles deep and extends all the way through the earth's crust to the mantle below (allowing magma to intrude upwards through the cracks), is actually called a *failed rift*, because, unlike other rifts that *did* split apart along the coasts, this rift did not completely split. If it had, ocean crust would have begun to form below it, and today the region might well be part of an ocean instead of landlocked as it is. The

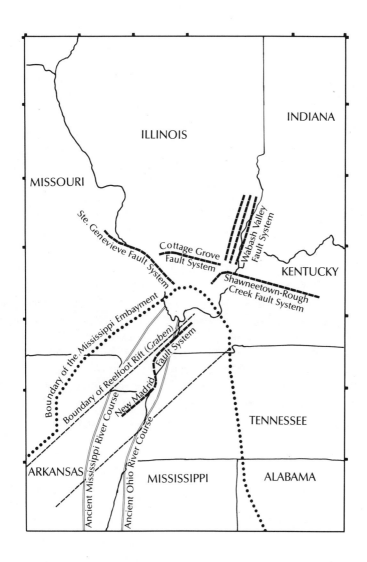

Map 2. Major Earthquake-Related Features in the Upper Mississippi Valley

Reelfoot Rift is actually the northern arm of a *trio* of rifts that met like spokes of a wheel a hundred miles or so north of present-day New Orleans. The two other arms (one extending southeast from this point, the other southwest) *did* rift, creating the Gulf of Mexico. Other failed rifts exist within the North American Plate, but none is as large as the Reelfoot Rift.

As the continent was attempting to break along the Reelfoot Rift, large sections of rock between the breaking points (cracks) along the rift dropped down a mile or so as the crust on either side pulled apart. These down-dropped blocks are referred to as *grabens*. As a result of these grabens, a depression called the Mississippi Embayment formed approximately two hunded million years ago from what is now the southern tip of Illinois to the Gulf of Mexico. This embayment, roughly eighty thousand square miles in size, is actually a long, relatively shallow trough (like an inverted V), narrow at its northernmost point and gradually widening until it reaches the Gulf's coastal plain. This embayment was part of the Gulf of Mexico at one time, the shores of which thus lapped up around the current confluence of the Mississippi and Ohio rivers. (Marine sediments were also thus deposited in the Mississippi Embayment.) After the Gulf of Mexico eventually receded and the Mississippi and·Ohio rivers formed and began draining the eastern part of the United States through the embayment, the two rivers flowed parallel to each other and met just north of the present-day site of New Orleans. It was only much later that some type of geological activity (possibly an earthquake) repositioned the rivers, cutting off the Ohio River at its current confluence with the Mississippi at the southern tip of Illinois (see map 2).

The multiple layers covering the Mississippi Embayment cause one of the major problems confronting scientists trying to study earthquakes in this area. The layers of sedimentary rock, sedimentary soil (called *alluvium*), and windblown silt (called *loess*) that blanket the embayment for between one and four miles in depth prevent easy location of the various fault lines and systems that lace the bedrock. (In California, there is little such covering. The San Andreas Fault is so close to the surface, one can literally stick a finger in it in places.) A second, related problem is that the soils and soft sedimentary rock fail to hold their profiles, thus eventually masking the results of earthquakes that have occurred in the past. In other words, since the soft rock formations are subject to various reformations and shifting over time, the geological records of fault movement are not preserved for very long, making it impossible for scientists who do dig into the embayment soils and rock to obtain accurate information on the effects of earthquakes that have occurred thousands of years in the past.

The soil and rock structure in the area, moreover, presents a problem that goes beyond making the study of Mississippi Embayment earthquakes difficult. Earthquakes that occur in many areas east of the Rocky Mountains tend to cause damage over a greater area than those of similar magnitude occurring west of the Rocky Mountains (along the San Andreas Fault, for example) because of differences in the physical properties of the rock in the upper crust (a layer of rock approximately fifteen miles thick). The upper crust east of the Rockies is a much more effective transmitter of an earthquake's destructive waves, spreading them over hundreds, even thousands, of square miles. Thus, while an earthquake of x magnitude occurring in California may cause damage to one or two counties, an earthquake of similar magnitude occurring in the Mississippi Embayment may cause damage to large sections of one or two *states*. This explains why the San Francisco earthquake of 1906, which was only slightly less powerful than the New Madrid earthquakes of 1811–1812, only caused damage to the region immediately surrounding the city, while the New Madrid shocks shook approximately two million square miles. Damage from an earthquake in the Mississippi Embayment, it is variously estimated, can be between twenty and one hundred times greater than from an earthquake of similar intensity occurring west of the Rockies.[12]

Mississippi Embayment soil structures, then, are destructive masks. While nothing can be done about the destructive element, scientists have been able to overcome the soil's masking qualities—at least partially. With modern technology, scientists have been able to locate and map underground fault structures and zones by sending sound waves deep into the earth and observing where the waves break. These breaks indicate fault structures. Coupled with historic information on where earthquakes have occurred in the area in the past, along with gravity and aeromagnetic maps of the region, scientists now have a very good initial understanding of where the most active faults, fault systems, and fault zones are in the region.

Though the upper Mississippi Valley is laced with a number of fault systems (among them the Cottage Grove, Rough Creek, Wabash Valley, and Ste. Genevieve systems), the most active, and potentially most powerful, fault zone in the Midwest is, as might be expected, centered under New Madrid, Missouri, where scientists have located a large graben, approximately 120 miles long and 40 miles wide, in the Reelfoot Rift. Running through the center of this graben is a large, jagged fault, possibly fifteen miles deep, extending from northeastern Arkansas to near Cairo, Illinois. This area is known as the New Madrid Seismic Zone.

Midwestern Earthquakes

As explained earlier in this chapter, the North American Plate is being slowly pushed westward (away from Europe and Africa) from its source in the middle of the Atlantic Ocean. This process began approximately two hundred million years ago, reversing the process that had pushed the plate eastward (toward Europe and Africa) for roughly three hundred million years and caused the original failed rift in what is now the Mississippi Embayment. Whereas the Midwest was being pulled apart five hundred million years ago, it is now being compressed—pushed westward from the east (where the North American Plate is forming in the mid-Atlantic) and eastward from the west (by the Pacific Plate, which is grating along the western edge of the North American Plate at the San Andreas Fault).

One theory suggests that as the North American Plate is moving over the hot mantle below it, part of the plate is dragging along the mantle, causing activity along midwestern faults, many of which extend down as far as the mantle and are capable of injecting the molten rock up through their cracks.[13] A second theory suggests that it is compression, discussed in the previous paragraph, that causes earthquakes in the region. Since the faulted zones are areas of crustal weakness, they are the areas likely to give first when the pressure from compression becomes too great.

An interesting theory proposed more than once argues that one giant, continuous—or almost continuous—rift or fault system runs from New Madrid along the Ohio and St. Lawrence rivers to the Atlantic Ocean near Newfoundland. Some of the most powerful earthquakes in the eastern part of the continent have occurred along this imaginary line (including a number of major shocks in Ohio and along the St. Lawrence River.) Most scientists dismiss this theory, noting that geological formations along the supposed pathway are too different and that while there are a number of patches of significant earthquake activity along the line, there are also areas that have exhibited no activity in recent times. Proponents of the theory, however, argue that seismic activity does not need to occur along all points of a rift or fault system in order for the rift or system to be continuous. In addition, they argue, these "quiet" areas could simply have been inactive for the past two hundred years, but could have been active before the country was settled. One of the first steps being taken to either prove or disprove the "megafault" theory is to try to connect the New Madrid Seismic Zone to a parallel active seismic zone northeast of it running through southwestern Indiana in a northeasterly direction. To date, these are considered two separate zones, separated by an inactive and geographically different wedge of land in western Kentucky and southern Illinois, but if research proves fruitful, these two sections may be found to be linked. Such research concerning other known systems along the "line" could eventually uncover such a massive rift from Arkansas to Newfoundland.[14]

4. Earthquake Prediction

Boom and Bust for Earthquake Prediction

A string of successful earthquake predictions in both the United States and China during the mid-1970s buoyed the scientific community, as successful earthquake prediction seemed to have become a reality. In August 1973, a small earthquake shook the Adirondack region of upstate New York. Scientists at Columbia University's Lamont-Doherty Geological Observatory had successfully predicted its time (two days in advance), location (Blue Mountain Lake), and magnitude (2.5).[1] On Thanksgiving Day, 1974, a small tremor shook the area near Hollister, California (just south of San Francisco). The night before, a group of geologists had predicted a tremor there within the week.[2] On February 4, 1975, at 2:00 A.M., Chinese officials ordered the ninety thousand residents of Haicheng, China, to evacuate the city and move into the frigid countryside. Later that afternoon, a violent shock measuring 7.4 on the Richter scale virtually destroyed the city, severely damaging 90 percent of its buildings, as the inhabitants watched in relative safety from the outskirts of the city. The successful prediction, it is estimated, saved ten thousand lives.

In 1976, University of South Carolina scientists successfully predicted a small tremor near Lake Jocassee in that state two days before it occurred.[3] On January 6, 1977, another small earthquake (magnitude 3.2) shook the area six miles from Hollister, California. A week earlier, six United States Geological Survey scientists had predicted a 3.5 tremor within nine miles of Hollister.[4] A 1978 earthquake off the southern coast of Mexico had been successfully predicted by a group of University of Texas scientists the year before. The quake was almost exactly the magnitude they had estimated and occurred just thirty miles from where they had predicted it would.[5] The euphoria was short-lived, however. A number of predictions proffered throughout the 1970s failed to materialize, and a number of fault zones being carefully monitored for preliminary signs of activity experienced earthquakes without any preliminary warnings. Just a year after China's successful prediction of the shock near Haicheng, an 8.2 magnitude earthquake ruined the city of Tangshan, killing an estimated eight hundred thousand people (as discussed in chapter 3). Some early warning signs of an earthquake in the region had been reported, and a preliminary warning had been given. When the signs failed to continue, however, observers let down their guard.[6]

Many conservative U.S. scientists love to point to this failure as proof that Chinese earthquake prediction techniques are not all they have been claimed to be. While U.S. scientists rely almost exclusively on possible precursors that can be measured with sensitive scientific instruments, the Chinese government, while employing similar methods, also has at its disposal a cadre of approximately one hundred thousand peasants scattered throughout the country near fault zones. These peasants have been instructed in how to recognize other, less conventional, possible earthquake precursors—the most common of which are changes in animal behavior and changes in water-well levels and water characteristics. Using this combination of scientific and "nonscientific" techniques, the Chinese successfully predicted thirteen earthquakes between 1974 and 1976, all of which resulted in evacuations and hundreds of thousands of lives saved. During another two-year period, however, only eighteen of thirty-one predictions proved accurate.[7]

Currently, the United States, China, Japan, and the Soviet Union are all actively engaged in earthquake prediction programs, but success is still elusive. Scientists have found that there are no consistent earthquake precursors: while some possible precursors may appear before some earthquakes, they do not appear before others. Another confounding problem is that earthquakes do not always occur after many of these supposed precursors do appear. The problem is two-fold, then. There are no known precursors that appear before all (or even most) earthquakes, and even when possible precursors do appear, there is no assurance of an ensuing event.

Earthquake Precursors

As discussed above, there are a number of possible signs of an impending earthquake, though not all earthquakes exhibit such signs (and many of the signs appear without being followed by an earthquake at all). One of the more common events thought to be an earthquake precursor is one or more *foreshocks* or "microearthquakes." There are a number of problems with this sign, however. First, not all earthquakes are preceded by foreshocks. In California, for instance, only one-fourth of the earthquakes are preceded by foreshocks. In Japan, only sixty of fifteen hundred earthquakes occurring between 1926 and 1961 were preceded by foreshocks.[8] Second, until a major earthquake occurs, it is difficult to tell if a foreshock is actually that—a foreshock—or an independent, weak earthquake (or series of microearthquakes) that will not be followed by a more sizable event. In other words, it is only *after* a major earthquake that scientists can determine if the preliminary shocks were actually foreshocks. Third, even if the foreshocks are precursors of a larger earthquake, they can come a day in advance or a year in advance, thus not being very helpful in predicting when a more powerful shock will occur. Finally, foreshocks do not necessarily assist in determining the magnitude of ensuing major events. Large earthquakes may be preceded by just a few weak microearthquakes, while smaller earthquakes may be preceded by a large number of substantial foreshocks. In defense of the predictive value of foreshocks, however, one can argue that they were one of the major signs the Chinese used to successfully predict the 1975 Haicheng earthquake.

Another precursor, *dilatancy*, had the scientific world quite excited in the mid-1970s. As discussed in chapter 3, prior to earthquakes, pressure increases on rocks, and as they are strained to their breaking points, they swell and dilate as small cracks form and begin to fill with water. Delicate instruments can often measure this expansion (or dilatancy) by measuring small bulges along the fault as well as changes in the velocity of underground seismic waves. This sign, however, also has not materialized over the years as being accurate, since instruments have not measured any of these changes prior to some powerful earthquakes *and* because such changes have not always resulted in earthquakes.

Changes in water level and purity often occur prior to earthquakes, but again, not always. As rocks crack and absorb water, groundwater levels can change, and the water can become cloudy. Almost two thousand years ago, the Roman scholar Pliny reported that cloudy, foul-smelling well water was an indication of an impending earthquake, and Japanese samurai were using the method to predict seismic events as far back as the 1700s. In Europe in 1808, four hours before an earthquake, and again in 1835 before an earthquake in Chile, women washing their clothes in nearby rivers were startled to find the water suddenly turn murky and begin flooding its banks.[9] In China, rural observers are instructed to report any significant changes in water-well levels, in the belief that such shifts can indicate impending seismic activity. Observers are also instructed to file reports anytime usually clear well water suddenly becomes cloudy or murky, churned or muddy, and/or gives off foul smells, which may come from underground gases. (Changes in well-water levels and purity were another sign the Chinese used to predict the 1975 earthquake.) Soviet scientists have also used this method to successfully predict some of their earthquakes.[10]

Changes in the electromagnetic field around a fault zone also often occur prior to earthquakes, again as a result of the cracking of rocks under pressure. As cracks open, electricity is allowed to flow more freely through the groundwater. As the cracks close just prior to the earthquake, electrical flow is decreased. Under laboratory conditions, rocks being fractured generate these same impulsive electromagnetic emissions. By feeding electric current through the ground between two points on a fault, scientists can detect any voltage changes that may occur. Other electromagnetic activity occurring along the fault can create atmospheric radio signals and cause nearby compasses to go haywire. For three to five days before the 1976 Tangshan earthquake, which the Chinese did not go on to correctly predict, many of the military and civilian radio-communication receivers in the area picked up unusual interference, later suspected to be related to the impending earthquake.[11]

More visible to the average observer may be *earthquake lights*. Although not fully accepted by the scientific community because they have not yet been observed sufficiently by scientists, earthquake lights appear as flashes of light coming from the ground near the epicenters of some earthquakes. In the distance, the lights often appear as bright luminescence covering the sky around the epicenter, similar to the effect of sheet lightning, but usually lasting for a longer period of time. Earthquake lights have been seen as far away as seventy miles.[12] Colors range from pale blue or white to reddish-orange. One of the few reports of a closeup view of earthquake lights comes from a California farmer who noticed flashes of light coming from a nearby hillside during a sequence of small earthquakes in 1961. Inspection of the hill disclosed no downed power lines or any other obvious source of the flashes.[13] Other reports of earthquake lights come from the 1811–1812 earthquake sequence in New Madrid (discussed in chapter 1), where residents reported flashes near the epicenters and observers many miles to the east reported the glowing in the western sky. While most sightings of earthquake lights occur during and/or shortly after earthquakes, people have reported seeing them prior to earthquakes, even up to several hours before the quakes. They, therefore, must be considered possible precursors. There are at least three possible causes of the lights: (1) the buildup and subsequent release of electromagnetism from the ground (as discussed above), (2) violent, low-level air oscillation resulting from the seismic activity, and

(3) release of methane gas from the ground (discussed more fully below).

One of the more intriguing precursors to earthquakes is the emanation of radon gas from the ground. Radon, a colorless, odorless, radioactive gas, is trapped within all rocks, and when rocks are fractured, as they are prior to earthquakes, the gas is released in greater quantities than usual. (Radon is always emanating from the ground in small quantities.) It is theorized that these excessive quantities of radon can work their way toward the surface, where they can be detected by sensitive strips of film buried a few feet underground. Radon has been used to successfully predict a few earthquakes in China, the Soviet Union, and the United States. In some cases, it has been the only precursor used to predict the activity successfully. In others, it has been simply one of many precursors observed and subsequently used to predict the seismic events. When it is successfully used to predict earthquakes, the increase in the level of radon emanation generally occurs four to six weeks prior to the earthquake it precedes.

As with all other precursors discovered and tested, however, radon emanation is not foolproof. Some earthquakes are preceded by increases in radon emanation; others seem not to be. In addition, large increases in the radon level can often be caused by other phenomena, such as changes in water level (caused by something other than seismic activity) and changes in barometric pressure, thus leading to false alarms. Using radon to predict earthquakes, therefore, has recently fallen into disfavor. Scientists now believe that large quantities of radon come to the surface as a result of something other than seismic activity.

There are other gases that may come to the surface prior to earthquakes. Among these are hydrogen, helium, carbon dioxide, and methane. For two months prior to the 1975 China earthquake, farm workers reported an extremely strong odor in the fields. On some days, it was so strong that it actually forced them to stop working. One worker observed gas bubbles bursting from a trench, and another lost consciousness. This information, too, was used by the Chinese to predict the February cataclysm.[14] In Japan, research shows a clear correlation between hydrogen released from the earth and seismic activity (fault movement). Studies on the release of helium from the ground in California are proving helpful, as scientists are coming up with relatively good percentages of prediction.[15]

Earthquake Control

In light of what is known about possible precursors to earthquakes, is there any hope for ever preventing major quakes? Indeed there is, but the efforts are fraught with unknowns and some very messy potential legal implications and complications.

The first indication that major earthquakes might be prevented by the controlled release of stored energy in the form of smaller quakes occurred in the late 1950s when the United States began underground testing of nuclear devices in Nevada. It seemed that the blasts were followed by a series of shallow-focus earthquakes along preexisting faults.[16] Exploding nuclear devices underground to shake loose a moderate earthquake before it builds up even more destructive energy and issues forth as a powerful shock, however, is something no one wants to take responsibility for. Any damage resulting either from the blast itself or the ensuing earthquake would certainly be subject to litigation, and a number of lives might be at stake.

The second indication also occurred by accident. In 1962, the U.S. Army began disposing of contaminated wastewater that had been used in the manufacture of nerve gas by injecting it deep into the ground near Denver, Colorado. Soon afterward, the area began to experience a number of earthquakes. Since the area had no historical record of seismicity at all, the earthquakes caught everyone's attention. Subsequent research suggested that the pressurized water had triggered the disturbances. The fact that no earthquakes occurred after the injection of wastewater ceased added support to the theory. Scientists came to the conclusion that the possibilities of the earthquakes occurring randomly so close in time and place to the disposal area was one in 2.5 million.[17] Three years later, in a completely unrelated event, an oil company began pumping pressurized water deep into the ground in an oil field in northwestern Colorado (near Rangely) in order to force more petroleum out of the sandstone. Within a year, a number of small earthquakes began occurring throughout the Rangely area. Scientists received permission from the company to set up seismometers in the region, and they discovered that the epicenters of the shocks were in the two areas where the injection pressures were the highest. As injection increased, seismic activity also increased.[18]

Little if any further research on triggering earthquakes has been conducted in recent years. For one thing, it would take at least one hundred earthquakes of magnitude 6.0 to release the equivalent energy of one magnitude 7.5 earthquake, and, again, the legal and moral implications may simply be too great for anyone to take responsibility for setting off an earthquake designed to relieve stress and prevent a more powerful event. In addition, since so little is known about the mechanics of artificially triggered earthquakes, there is no way to know for sure how powerful an artificial event might be. It is theoretically possible that pumping water under pressure into the ground along a fault could trigger an extremely powerful earthquake that would take hundreds of lives and cause hundreds of millions of dollars worth of damage. Such an earthquake, if allowed to occur naturally, might not happen for a hundred or more years, by which time better methods of prediction, prevention, and mitigation might be available.

Earthquake Cycles?

While it is well accepted in scientific circles that earthquakes are caused by the sudden release of a gradual buildup of stress along a fault, no one is sure if there is a final triggering mechanism that causes the sudden release. Well back into the 1800s, though, even before we understood earthquake mechanics, scientists were attempting to make correlations between the frequency of earthquakes and any number of periodic or anomalous events:

comets, meteoric activity, sunspots, solar flares, polar wobble, lunar positions, planetary positions, seasons, and hours, among them. In brief, some studies have found strong correlations, others have found none, and still others have found some.[19] Current wisdom seems to suggest that any (or maybe all) of these events may have some kind of effect on earthquakes. It is thought by those who pursue these theories that the energy released or caused by these events may be the "straw that broke the camel's back." In other words, if enough energy is stored in a certain location to precipitate an earthquake, one of these periodic events might be just the extra force needed to cause the earthquake. Naturally, it is not suggested that these periodic events cause earthquakes. It is simply suggested that these events may precipitate the sudden release of energy, not that they are powerful enough to cause stress to build up.

Going back centuries, folklore has held that comets are harbingers of disaster. When comets are seen in the sky, it has been suggested, the world suffers all sorts of calamities, not the least of which are earthquakes. Much of the folklore surrounding the 1811–1812 earthquakes, in fact, revolves around the comet seen rocketing through the sky during the months preceding the shocks. The most popular comet, Halley's, has not seemed to have much effect on seismic activity, however. Records do not indicate increased seismic activity during the comet's 1835 and 1910 visits to the planet, though there has been a flurry of recent earthquake activity worldwide during, but especially following, the comet's most recent visit (1985).

Another celestial event that many observers claim is seen just prior to earthquakes is a meteor or meteor shower.[20] Just before the disastrous 1663 St. Lawrence River earthquake in Canada observers reported seeing meteorlike luminous objects in the air.[21] At least a dozen other reports of meteor activity prior to earthquakes have been written about. One theory suggests that people are actually seeing earthquake lights, a second that meteors crashing to the ground themselves actually trigger earthquakes. Neither theory has been proven or disproven. (It is also interesting to note dozens of reports of increased UFO activity just prior to earthquakes. While those with vivid imaginations may be prone to think extraterrestrial aliens are responsible for our seismic shocks, it is more likely that people are instead seeing earthquake lights.)

The theory that the sun can precipitate earthquakes goes back to the 1800s, when some scientists attempted to correlate seismic activity with solar activity (which occurs in approximately eleven-year cycles).[22] Most of the studies in recent years have found a relatively strong correlation between solar activity (solar flares and sunspots) and earthquake activity. A solar flare releases energy equivalent to ten to one hundred *billion* one-megaton hydrogen bombs, which, if harnessed, would be enough power to supply the United States with energy for thousands of years. The flares also release large amounts of radiation, electrical and magnetic fields, and high-energy particles that are the cause of increased geomagnetic storms on earth as well as direct electrical surges in power lines and even the ground itself. It is therefore not totally unlikely that such activity could trigger earthquakes.[23]

Another periodic event some scientists feel is connected to earthquake activity is Chandler's Wobble, the name given to the periodic *nutation* (wobbling) of the earth's poles. Occasionally, some solid correlations between earthquakes and the Wobble are presented, but little has been discovered relating to cause. Some scientists suggest Chandler's Wobble triggers earthquakes. Others say earthquakes trigger Chandler's Wobble. Still others claim the two are caused by some third event, possibly the release of elastic energy stored in the earth.[24]

The moon has also been suggested as a triggering mechanism for earthquakes. Studies have attempted since the nineteenth century to correlate the moon's position with seismic activity. Some scientists have suggested that tremors are more common during full moons and new moons than at other times of the month. (Some studies bear this relationship out; others dispute it.) Other scientists have suggested that it is rather the distance of the moon from the earth, not the position, that triggers seismic events. A number of studies have found slight to moderate increases in earthquake activity during perigee (when the moon is nearest the earth), with less activity during apogee (when the moon is farthest from the earth). There are some reasonable explanations for the moon's possible effect on seismic activity. The moon, of course, affects the oceans' tides, a result of the changing gravitational pull of the moon in relation to the sun and the earth. The earth, though, has tides of its own—"solid-earth tides"—also triggered by the moon. It is hypothesized that the effects of these slight gravitational stresses on the earth—the ups and downs, expansions, contractions, and compressions—can trigger seismic activity. While bodies of water easily respond (shift) as a result of the pressures on them, the earth, with roughly the strength of a steel ball, does not respond as easily. It is therefore quite possible that solid-earth tides can trigger activity at weak points in the crust—fault zones.[25]

The theory that earthquakes have an *annual* periodicity has also received a great deal of attention. As early as the 1800s, some scientists found very high correlations between earthquakes and cold weather, with the greatest number of earthquakes per month occurring between November and February in the Northern Hemisphere.[26] No satisfactory explanations have yet been offered why this seems to be the case.

The Gap Theory

Although not directly related to periodic events that themselves trigger earthquakes, the "gap theory" nonetheless relates to earthquake periodicity. The gap theory suggests that earthquakes occur along fault lines on a somewhat regular spatial basis. As activity is recorded along a line, it becomes clear that certain sections of the line have been inactive for a rather long period of time. The gap theory suggests that these inactive sections, or "gaps," are likely to be the locations of the next earthquakes, since they may be areas where seismic energy is being stored without being released. (One must be careful, however, because an aseismic zone, or gap, can also exist along a section of a fault that is experiencing "creep"—that is, where no energy is being stored.) Few scientists today question the validity of the gap theory. Most respected seismologists and geophysicists agree that the gap theory has a solid record of success in terms of prediction.[27] The September 1985 earthquake in Mexico, for instance, was expected, since that segment of the fault line along Mexico's western coast had been inactive longer than any of the other segments.

The major problem with the gap theory is that its "time windows" are generally too large for useful prediction. Using the gap theory, scientists can predict seismic activity in a certain section of a fault within ten or twenty years, but rarely any more accurately. Coupled with other indicators of increased strain (as discussed in the previous section), however, the gap theory may prove very useful in helping to pinpoint volatile locations so they can be carefully monitored. The current value of the gap theory, in other words, is that it helps to isolate where future seismic activity should occur. (Knowing the location is of great value, because scientists can then concentrate their monitoring efforts in those areas, instead of having to guess where the next activity might occur.) Other methods need to be perfected to help predict more accurately when the activity will occur.

One of the most recent areas of research in earthquake prediction has been the study of the possibility that worldwide earthquakes are connected; that seismic activity caused by plate movement in one section of the world may trigger seismic activity elsewhere. However, nothing definitive has come of this research yet.

Animal Earthquake Prediction

The historian Diodorus reported that in 373 B.C. animals in the city of Helice, Greece, including rats, snakes, weasels, and other burrowing creatures, left the city in droves along a connecting road toward the city of Koria. Five days later, Helice was destroyed by an earthquake and swallowed by the sea. On February 5, 1783, in Messina, Italy, geese cackled uncontrollably, and dogs howled so incessantly that an order was given to shoot them. Soon afterwards, a large earthquake hit the nearby town of Calabria. On July 25, 1805, huge swarms of locusts crawled through the streets of Naples, Italy, toward the sea. The following day, domestic animals became unmanageable and tried to break out of their stalls and corrals. Shortly afterward, a powerful earthquake struck the city. On February 20, 1835, residents of Talcahuano, Chile, remarked to one another with surprise about the large flocks of sea fowl leaving the coast and flying inland. Dogs left the city shortly afterward. About noon, Talcahuano was leveled by an earthquake. It is believed every animal left the city before the shock.[28]

On April 17, 1906, dogs in San Francisco, California, barked and howled all night. The following day, cows fled in panic, and twenty-five to thirty horses tore loose from their stables. A dog ran about wildly and jumped out a first-story window. A few moments later, the earthquake struck the city. On June 27, 1966, rattlesnakes fled the hillsides around Parkfield, California, and invaded the city. The migration occurred for two days. On June 29, a strong tremor rumbled through the area. On February 8, 1971, at least two police patrols reported independently to their dispatcher that large numbers of rats were scurrying through the streets of San Fernando, California, a phenomenon they had never witnessed before. The following day, a powerful earthquake shattered the city. On February 2, 1975, twenty piglets in Haicheng, China, cried wildly in their pens—most had had their tails bitten off and eaten—and fish jumped out of a pond onto land. The following day, February 3, homing pigeons left the city and normally docile cattle locked horns and began to fight. Birds in the city picked up eggs from their nests and flew away. Rats wandered through the streets. On the morning of February 4, a frog was discovered frozen outside its hibernating hole, and dogs began barking wildly and digging holes in the ground. Taking these and literally hundreds of other reports of agitated animal activity into account, officials ordered the evacuation of the city. Hours later, residents experienced the powerful earthquake from the hillsides, and Haicheng crumbled.

The evidence of animal prescience of earthquakes is simply too overwhelming to ignore. There are hundreds of thousands of reports of uncharacteristic animal behavior prior to earthquakes well into ancient times. While many are reported after the fact and many more are not all that unusual, a number of such accounts are anomalous enough to lead many researchers to believe that animals sense some, but not all, earthquakes.

The Chinese have been studying strange animal behavior as precursors to earthquakes for three thousand years and have had a number of successes (at least fifteen) predicting seismic activity as a result.[29] Some estimates suggest their predictions have saved perhaps a million lives.[30] Prior to the 1976 Tangshan earthquake, an earthquake they did not predict in time, officials collected almost twenty-one hundred reports of anomalous animal activity but did not gather enough other information in time to order an evacuation of the city.[31]

In Japan, animal observation has also had a long and interesting history. Since most of the island's tremors originate just offshore, most observations involve anomalous fish activity. However, the January 14, 1978, earthquake was preceded by well over one hundred reports of strange activity involving mammals, birds, insects, reptiles, amphibians, as well as fish.[32]

Although not nearly as common, there have also been reports of humans experiencing strange sensations prior to seismic events. Records from Italy and other countries report large numbers of people suffering from nervous irritability, restlessness, tremors, general malaise, nausea, breathing difficulties, dizziness, inclinations to vomit, and heart palpitations prior to earthquakes. For two months prior to the 1948 Soviet Union earthquake, many residents complained of heart problems, but medical exams showed no signs of any disorders. Following the earthquake, all heart complaints ceased.[33]

One possible explanation of human complaints prior to earthquakes is that they are simply reacting to anomalous animal activity. People may well experience stress as a result of their pets and/or livestock acting so unusually. There are, however, some credible theories on what triggers animal activity, and none of these precludes the possibility that rather sensitive people respond to the same triggering mechanisms as other animals.

One possible mechanism is that animals hear the underground sounds that occur prior to earthquakes—the result of seismic activity leading to the actual events. Another possibility is that they respond to the low-level electromagnetic activity and changing atmospheric ionization that precede earthquakes. A third possibility is that they sense one or more of the gases (radon and methane, for example) that often emanate from the ground prior to the shocks.

Little rigorous research has been conducted on animal behavior before earthquakes; even less on why it may occur. It may be years before scientists understand the mechanism(s) of such behavior. Some formal research *is* taking place, however. In California, experiments were begun in the 1970s to study animal behavior prior to earthquakes along the San Andreas Fault. Scientists observed the behavior of caged rats, mice, and cockroaches prior to seismic events. In the early 1980s, a group of scientists set up a network of well over one thousand volunteers to phone in any anomalous animal activity as it occurred. Using the latter method, researchers were able to predict seven of thirteen earthquakes in a four-year period. The experiment has not been an unqualified success, though, since most of the calls came from people after the earthquakes had taken place, claiming that they had noticed strange animal behavior before the events but had simply failed to call them in. These reports, of course, were deemed invalid.[34] Another researcher uses a similar "hotline" program asking volunteers to report any unusual physical symptoms they experience (flulike symptoms such as chills, hot flashes, nausea, headaches, and fatigue, for example). Using this information, she is reported to have predicted 75 percent of California's recent earthquakes. One of her volunteers has only called four times to report symptoms but has been on the mark 100 percent of the time. After each call, an earthquake of moderate intensity has struck the area.[35]

In spite of some moderate and scattered success, U.S. research into anomalous animal activity does not seem to have a particularly active future. For one reason, successes have not been startling enough to generate much excitement among the traditionally conservative U.S. scientific community. European and Asian scientists tend to move into untested waters much more readily than their counterparts in the United States. Only when foreign research begins to show solid results do U.S. scientists become more than slightly interested and shed their skepticism. (This is true in many areas besides earthquake prediction, medical research and biometeorology—the study of the effects of atmospheric conditions on humans and animals—among them.) Indications seem to point to the realization that until foreign scientists determine the *cause* of preseismic animal behavior, such research—always hampered by a lack of available funding—will progress at a snail's pace in this nation.

Earthquake Prediction for the Midwest

The implications of worldwide earthquake prediction techniques for the Midwest, unfortunately, are very limited at this point. There are a number of problems with using currently known prediction theories for prediction of seismic activity in the Mississippi Valley. First, the known history of tremors in the region—only about three hundred years—is not long enough to confidently establish recurrence times. That earthquakes in the Midwest, while often more devastating than those in California, occur between five to seven times less frequently than their West Coast counterparts only adds to this problem and makes establishing reliable recurrence times even more difficult. As a result, even the most promising earthquake prediction technique, the gap theory, has not yet been proven valid in the region. "Even though the theory is promising, there is no indication it will hold up in the Midwest," admits Dr. Otto W. Nuttli, professor of geophysics at St. Louis University (St. Louis, Missouri) and the world's leading expert on midwestern seismicity. "There are what appear to be gaps, but we've had no earthquakes yet to fill in those gaps. With the recurrence times being so much longer with intraplate earthquakes, we'll simply have to wait to see if the gap theory will hold true in the Midwest."[36]

Another problem is that, again, the faults in the area are buried under miles of sediment, so it is not just a matter of setting up instruments along and within a fault system as is done in California, where the San Andreas Fault is on the surface. To monitor a fault in the Midwest requires drilling deep into the earth. As a consequence, there is really little opportunity to measure the effects of strain buildup along the New Madrid Fault System or any other fault system in the region.

To date, about the only method of generalized prediction being used for Mississippi Valley earthquakes is estimated recurrence time. (The radon theory has been tested somewhat successfully in the New Madrid Seismic Zone, but while results seemed to indicate positive correlations, most experts caution against too much optimism.[37]) Scientists have arrived at data for such predictions in two ways: (1) using physics and mathematics, St. Louis University's Otto Nuttli has spent a great deal of time studying earthquake energy, and by extrapolating the rate of occurrence of present-day smaller earthquakes has arrived at estimated recurrence times for larger earthquakes; (2) by digging a trench in the New Madrid Seismic Zone, the United States Geological Survey (Denver) has found indications of episodes of faulting and sand blows occurring previous to the 1811–1812 New Madrid earthquakes.

It is encouraging that the studies independently came up with similar results. Based on his research, Nuttli estimates that large earthquakes (those similar to the 1811–12 tremors) occur on average every six hundred to one thousand years.[38] The USGS, in its field research, found two episodes of faulting and sand blows in the two-thousand-year period before the 1811–1812 tremors, indicating that such large earthquakes occur on average approximately every six hundred to seven hundred years.[39] A USGS professional workshop held in 1980, which brought together well-known experts on midwestern seismicity, reported general agreement with these figures, though those attending refined the estimates slightly. According to the final report, major earthquakes within the New Madrid Seismic Zone are estimated to occur every six hundred to seven hundred years, whereas moderate earthquakes within the other Mississippi Valley earthquake zones (the Ste. Genevieve, Cottage Grove, Wabash Valley, and other zones) are estimated to occur every one thousand years.[40] (There is no evidence at this point that major earthquakes could occur in these other zones. Only the New Madrid Seismic Zone seems capable of major earthquakes, at least to date.)

Moderate earthquakes (6.0 to 6.5 magnitude) within the New Madrid Seismic Zone, such as occurred in 1843 and 1895, recur an average of once every seventy-five to one hundred years, according to the best estimates.[41] The area, therefore, is very much due for an earthquake with a magnitude of at least 6.0, since the last one of this magnitude occurred over ninety years ago (in 1895).

A quick assessment of the estimates for the next major earthquake may lead us to believe that we are safe until at least the year 2400 (approximately six hundred years after the 1811–1812 earthquakes). This is a dangerous assumption, though, for at least two reasons. First, the key word in the estimates given above is *average*. Average does not mean exactly, or even approximately, every six hundred to seven hundred years. Field work, again, found two episodes of earthquake activity within a two-thousand-year period. There is no indication they were evenly spaced six hundred to seven hundred years apart. In China, for instance, where accurate records go back as far as three thousand years, major earthquakes in certain regions have occurred one right after another, with the faults then remaining inactive for hundreds of years. Some earthquake zones can have a great deal of activity for a number of centuries, then little activity for hundreds of more years. It is impossible, therefore, to say that the New Madrid Seismic Zone has had, or will have, major earthquakes every six hundred to seven hundred years. Earthquakes in the zone may be bunched all together; then the zone may be quiet for many centuries. In theory, then, the New Madrid Seismic Zone could erupt with a major, catastrophic earthquake at any time. In fact, Nuttli reports that his research indicates there may currently be enough energy stored in the fault system to trigger a quake of 7.6 magnitude if all the energy were released at once, which is not an impossibility.[42]

The second reason that feelings of present safety are dangerous is that (1) since the area is so heavily faulted, (2) since all the faults may not have been discovered yet because of their being buried under miles of sediment, and (3) since the known seismic history of the region is so short, there is no proof that another fault system in the area (either one of those known or one that has not yet been discovered) may not suddenly erupt with cataclysmic results. In other words, the fact that no other major earthquakes are known to have occurred in the area except the 1811–1812 New Madrid sequence does not mean there are no other fault systems in the area capable of producing very large earthquakes—fault systems that may have been lying inactive for centuries. In China, for instance, major earthquakes have unexpectedly occurred along fault lines that were previously thought to be capable of producing only small earthquakes.[43]

In theory, then, what lies below the region is largely unknown, as is what it is capable of doing. Even though an earthquake of 1811–1812 proportions is not expected in the area for a long time, the region simply cannot ignore the possibility that such an earthquake can happen at any time.

Are there any *specific* indicators to further help predict when a moderate or large earthquake will occur in the region? There does not seem to be any correlations between the frequency of Mississippi Valley earthquakes and the appearances of comets, although it is interesting to note that two of the most noteworthy and talked about comets of the nineteenth century were the Great Comet of 1811 and the Great Comet of 1843, comets that preceded the major Mississippi Valley earthquakes of 1811–1812 and 1843 respectively. The relationship is highly speculative, though, since comets are so common to begin with. Researchers may or may not find such a relationship some day. Preliminary research also seems to show no correlation between lunar phases and earthquake activity in the Mississippi Valley and no correlation between Chandler's Wobble and earthquake activity.

There seems to be some moderate correlations between regional tremors and solar activity, however. When the twenty strongest earthquakes (magnitude 5.0 and over) occurring in the Mississippi Valley since 1800 are plotted on a sunspot activity graph (as on chart 1), the following relationships can be seen:

● Eighteen of twenty of the earthquakes occurred during a sunspot spike that reached one hundred or less. Only two occurred during a large spike.

● Fifteen of twenty occurred within two years after a change in the direction of sunspot activity (that is, within two years after a spike began to rise or fall), including five that were *on* a spike shift—those of 1843, 1883, 1905, 1917, and 1923.

● Thirteen of twenty occurred when sunspot activity was at an average of fifty (which is the average of sunspot activity since 1800) or less.

● Six of the seven earthquakes that occurred above the fifty mark—those of 1838, 1883, 1905, 1917, 1927, and 1968—occurred within one year of the spike shift.

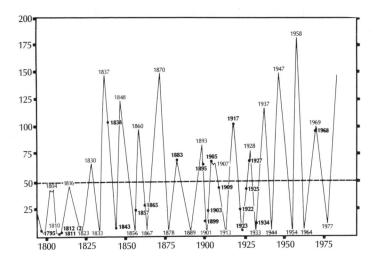

Chart 1. Sunspot Activity and the Most Powerful Mississippi Valley Earthquakes (Magnitude 5.0 and Greater), 1800–1980

Again, the possible relationships between sunspot activity and midwestern earthquakes is speculative. In fact, the 1895 earthquake, one of the five most powerful in the region since 1800, seems to be an anomaly in terms of sunspot activity; it simply does not fit any such pattern (other than being on a short spike). Since sunspot activity is known to affect geomagnetic activity on earth, however, there is no reason to believe that a more solid correlation may not be found some day for at least some types of earthquakes on the planet. Finally, it is most interesting to note that the 1811–1812 tremors, the most powerful in the Midwest's recorded history, occurred after almost two solid years of limited to nonexistent sunspot activity. In fact, there was absolutely no sunspot activity in the year 1810, a phenomenon that had not occurred since the early 1700s and has not occurred since 1810. The three-year period between 1809 and 1811 was characterized by the lowest consistent level of sunspot activity in the last three hundred years. If sunspot activity were to ever become that low again, the region might do well to at least consider some earthquake preparedness activity.

Another interesting correlation, and a much stronger one in fact, is that between moderate to strong regional tremors and the time of the year. As can be seen in chart 2, the moderate to strong earthquakes in the region occur much more frequently in the colder, drier months (eighty-seven total episodes between September and February versus forty-five episodes between March and August). More specifically, they occur almost three times as often during November, December, and January (forty-eight total episodes) as during May, June, and July (seventeen total episodes).

Midwest earthquakes cannot be predicted, but it initially seems that the moderate and powerful ones do occur in the colder months of the year and may occur just after changes in sunspot activity, especially just after periods of extremely low sunspot activity.

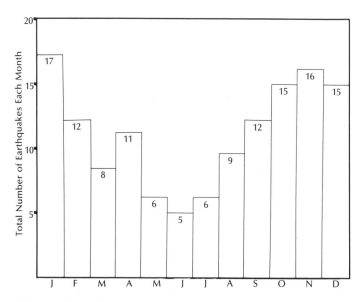

Chart 2. Distribution by Month of the 132 Most Powerful Mississippi Valley Earthquakes (Intensity V and Greater), 1795–1980

The Effects of Earthquake Prediction

There is no question that accurate earthquake prediction can save lives and property. With advance knowledge of tremors, citizens can make plans for orderly evacuation or at least prepare for the jolts (by taking precautions to prevent fires, by removing heavy objects from high places, and so forth); communities can, among other things, prepare emergency vehicles (by parking them outside structures so they are not buried under rubble, for example), lower water levels in reservoirs to help prevent dams from breaking, and prepare emergency shelters; and private enterprise can shut down such potential hazards as gas and oil pipelines and nuclear power plants. Predictability may not, however, be totally desirable, according to some.

Human Reactions to Earthquake Prediction

The late 1960s saw consternation growing in California. Much had been written about a major earthquake occurring along the

San Andreas Fault and sending California (or at least the western portion of it) into the Pacific Ocean. Local clairvoyants pinpointed April 1969 as the month when the "superquake" would occur. A group of Los Angeles's "flower children," who claimed they had been warned of the impending disaster by the Hopi Indians, fled into the desert. Pentecostal preachers saw that the time was right for some fire and brimstone sermons and led their flocks out of the state. April 1969 came and went. California remained.

An earthquake prediction in Peru in 1980 had different consequences. This prediction was made by two scientists, one from the United States Bureau of Mines and one from the United States Geological Survey, although the latter retracted his prediction prior to the time the event was to have occurred. Many people panicked; flights out of the country were booked solid, and emergency organizations prepared body bags. The earthquake never occurred.

Human reaction to earthquake prediction in China, however, is rather orderly. When an evacuation warning is given, citizens pack necessary belongings and leave their homes to wait out the expected tremor—regardless of the weather and regardless of how many times they have been inconvenienced by false alarms in the past. Social, political, and economic conditions are different in China, though, than they are in other countries. For one thing, homes and buildings in most of the communities are very poorly constructed, and people know death or serious injury is almost certain if earthquakes shake these structures. For another, the Chinese government actively seeks the participation of its citizens in earthquake prediction. Community members, thus, contribute to the predictions themselves; they are not simply the recipients of warnings from officials. For yet another, the political climate in China is such that citizens are taught to comply with governmental orders without much question. Finally, since the free enterprise system has been virtually nonexistent in the nation since 1949, no one has been overly concerned about loss of business as a result of an evacuation order. (It will be interesting to see if community reaction to evacuation orders will differ now that capitalism is beginning to gain a foothold in the country.)

Most professionals who specialize in disaster sociology suggest that the reactions of U.S. citizens faced with an authoritative earthquake prediction would be similar to their reactions to a hurricane prediction. If this is the case, reactions would vary widely, but the four most common types of reactions would be (1) compliance, (2) denial, (3) exhilaration, and (4) hysterics—with compliance being the most common reaction. The majority of people would, that is, comply with the directions of authority figures. A second group would try to deny the fact that a disaster was impending. They would tell themselves, "If disaster strikes, it will strike the other guy, not me." A third group would become invigorated. To these people, disasters are exhilarating. They are liable to throw a "disaster party" and try to "ride it out." In fact, not only do such people remain in their communities during disasters; they often become curiosity seekers and make special trips to other communities to view disasters firsthand. Such people feel in a sense that they are invincible. They feel nothing will touch them. A fourth group would feel that the disaster was aimed directly at them. As soon as a warning was sounded, they would fall apart and go into hysterics.[44]

Such were the reactions to the warnings for Hurricane Camille, which struck the Gulf of Mexico in 1969. While seventy-five thousand people calmly evacuated their homes and businesses along the Gulf, approximately twenty-five thousand people remained. Research indicates that those who remained tended to be from the upper middle class and were the best educated of the population. Many of them, however, admitted to prior ignorance of the power of the hurricane. Most of them felt they would not really be in danger. Others, especially the elderly, expressed a sense of fatalism—if their homes were going to be destroyed, they (the people) might as well be destroyed along with them. (Over one-third of those who died as a result of Camille—44 out of 123—were over sixty years old.[45])

Hurricanes are not earthquakes, of course, so it remains to be seen what people will do if scientists eventually learn how to predict earthquakes with a meaningful degree of accuracy and a genuine warning is sounded. One difference is that hurricanes are visible; earthquakes are not. People may tend to feel more in control of their safety during hurricanes, because they can actually see the danger coming. A second difference is that it is obvious to everyone when a hurricane has passed and is no longer a threat. If a hurricane warning is sounded and people evacuate, they know they will be able to return home within a specified period of time. If an earthquake warning were given, however, and the earthquake did not occur within a specified time frame, the potential would still be there. How long would those who evacuated remain away from their homes waiting for an earthquake to occur? A day? A week? A month? A year? And even after an initial shock, there is still the potential for devastating aftershocks.

At any rate, a number of elements must be present in the original warning for there to be a successful evacuation of an area in the first place. First, the warning must actually be a warning. It must convey a sense of urgency, or people will tend to ignore it as simply a bit of trivial scientific information. Second, the warning must be specific as to time, place, intensity, and general degree of risk. Third, it must be more than just a warning. It must be accompanied by specific instructions or recommendations for action. (For example, tornado warnings do more than tell people a tornado is spotted and on the way. The warnings add that people should take shelter and recommend where and when they should take such shelter.) Fourth, the warning must be promulgated by sources that people view as official (such as governmental officials and scientists), not by those with little credibility (amateur scientists, psychics, and astrologers, for example). Similarly, the media chosen to convey the message must be reputable. (Generally, people have faith that their local newspapers and radio and television stations receive their information from reliable sources.) Sixth, the message must be redundant— repeated over and over again to convey a sense of urgency and immediacy. Finally, the message must be consistent. If warnings are different, they create confusion. Generally, if good disaster warnings are provided and are accompanied by instructions on what to do, most people will follow the instructions.[46]

For a warning system to continue to be effective, there must be a solid success rate. If there are too many false alarms, people will simply tire of the warnings and eventually ignore them. A prime example of this phenomenon is the community warning siren system in this nation. The sirens are often sounded by mistake, and most people realize they will look foolish if they respond, unless the potential for disaster seems very obvious, such as is the case during a violent thunderstorm when people might expect a tornado. When the sirens do go off by mistake, though, police and fire departments are flooded with calls by a small percentage of anxious citizens. Naturally, this action defeats the purpose of the siren system. In a real disaster, it would probably be too late to take shelter after placing a call, and such calls tie up the lines of emergency service organizations when the lines are needed most for emergency work.

While the United States has no actual experience with earthquake warnings, some insight into human reaction can be obtained from surveys conducted in California in the 1970s. One study revealed that while 76 percent of the population said they would stockpile supplies if a credible warning were given for an impending earthquake, just 21 percent said they would leave the area even temporarily, most of them using the warning as an opportunity to take vacations. No one surveyed said they would leave the area permanently.[47] Another California study, however, suggested that approximately 60 percent of the residents of an area to be affected by an earthquake said they would evacuate if the event were officially pinpointed to within a week. Of these, 10 percent said they would move out of the area permanently. Those who planned to remain in the area said they would sleep outside at night and avoid older, taller buildings that might collapse.[48]

Moral and Legal Implications of Earthquake Prediction

If even generalized earthquake prediction becomes a reality and begins to tally up a series of successes, there could be serious implications for the economic viability of the areas where such events are predicted. Assuming a major earthquake is predicted for an area within two years, for instance, the following scenario might not be unlikely: Earthquake insurance becomes unavailable as insurance companies become concerned about whether or not they could meet the losses of such a catastrophe. Property values begin to decline. People begin saving money instead of making major purchases of durable goods, such as appliances and automobiles. Consequently, tax revenue in the community is cut from both property and sales. As a further result, a number of public services are cut. Businesses shut down and move out of the area. New businesses that have been considering moving into the area decide otherwise. Residents begin to move. People who have been considering moving into the area decide against such a move, and new construction begins to fall off. A number of retailers are forced out of business. Unemployment increases. Tourism, if it is an important part of the region's economy, dries up almost completely. Conceivably, the community is in greater ruin before the predicted earthquake than it might otherwise be afterwards. Even if the predicted event does not occur as expected, the community can be permanently crippled.

If earthquake prediction were to become a reality, the immediate impact would not be so much on the private citizen, as on the industrial/commercial sector. If a major earthquake were predicted in a community within even five years, such a prediction would affect new investments, new buildings, and new loans. Banks and other businesses and industries would certainly take notice. If there were a 50 to 80 percent probability of a large earthquake in a community within two years, social and economic systems might begin to adapt to that threat. If the prediction were 80 percent within two years and 100 percent within three years, there would most certainly be an economic downturn in the area.

There are, of course, a number of other problems the scientific community is currently addressing as it progresses toward earthquake prediction. Who, for example, will be responsible for predicting an earthquake? Who will be responsible for actually issuing the warning? Will the media provide a balanced assessment of the situation—on the one hand providing adequate warning, on the other preventing panic? Who will be responsible for helping the community to prepare? What responsibilities will these individuals, organizations, or governmental bodies have for the socioeconomic effects of their predictions and warnings—especially if they are false alarms?

Even though accurate prediction technology is not yet within the reach of science, scientists and governmental officials are already expressing concern over personal culpability. If an earthquake could be predicted with relative accuracy, would scientists have a personal and moral responsibility to make their individual predictions public? Could they be held accountable for the deaths, injuries, and damage that might result from an earthquake they privately predict but do not publicly announce? On the other hand, could they be held accountable for the socioeconomic downturn that would inevitably result if they made public a prediction for an earthquake that did not materialize within the estimated time frame? Some scientists are suggesting that public predictions should be withheld until the technical elements of prediction are capable of providing 100 percent accuracy.

There are other questions:

• Would the media give a prediction ample coverage to fairly warn the public, even though such reporting would inevitably lead to the partial demise of the area's socioeconomic structure, and thus of the media's own economic viability? (Newspapers and radio and television stations that warned of serious destruction in an area would certainly run the risk of losing subscribers and advertisers as people and businesses left the area.) Could the media be held culpable if they failed to provide adequate warning and an earthquake did indeed occur?

• If an earthquake were predicted for a community, say, within a week, would commercial building owners be obligated to close down for that week? If they did not close down, would they be liable for any deaths or injuries to people in or near their structures if they collapsed?

- Could local governments and/or their individual officials be held responsible for failing to move decisively enough to mitigate the effects of a predicted earthquake?
- Could regional consortiums set up with the purpose of helping their areas prepare for earthquakes be held liable for failing to carry out their charters?

While earthquake prediction in China is free from such litigious machinations, such prediction most certainly is not in the United States. "In our litigious society, little allowance is made for normal human errors if they have a substantial effect on someone's pocketbook or health," states a 1981 editorial on the subject of earthquake prediction in a scientific journal. "Unless allowances are made for the possibility of an erroneous or ill-timed warning, few scientists will be willing to risk the . . . litigation that could arise from open discussion," the editorial continues. "Unless some legal protection is forthcoming, the matter will most likely be resolved by the scientists retreating to an uninformative conservatism when asked to interpret their results in the public forum."[49] All of the other individuals, organizations, and governmental bodies that might grapple with the ethical issues involved in earthquake prediction might also find themselves grappling with the issues in a court of law—something none of them would relish.

While the scientific community continues its technical progress toward earthquake prediction, they and others tangentially involved must also work on the socioeconomic issues of prediction so they will be ready to make the appropriate decisions when the technology is available.

5. Earthquake Preparedness

Preparedness Efforts Around the Nation

The National Effort

The first serious attempt by the federal government to deal with the earthquake problem in the United States was the Earthquake Hazards Reduction Act of 1977 and the National Earthquake Hazards Reduction Program (NEHRP) that it created. In summary, the legislation was designed to (1) ensure informed coordination of land use, building design, public information, insurance, and warning and relief activities by federal, state, and local agencies; and (2) encourage research into earthquake prediction, the causes of earthquakes, zoning guidelines, seismic risk analysis for emergency planning, earthquake mitigation techniques for human-made structures, and socioeconomic adjustments to earthquake preparedness.

Initially, responsibility for the program was given to the U.S. Office of Science and Technology Policy of the Executive Office of the President, but in 1979 primary responsibility was transferred to the newly created Federal Emergency Management Agency (FEMA), with assistance to be provided by a number of other federal agencies, principally the United States Geological Survey (USGS), the National Science Foundation (NSF), and the National Bureau of Standards (NBS). (FEMA was created in 1979 by the merging of the National Fire Prevention and Control Administration, the Federal Insurance Administration, the Civil

Defense Preparedness Agency, the Federal Disaster Assistance Adminstration, and the Federal Preparedness Agency.) NEHRP recently developed a five-year plan (through 1989) for hazard delineation and assessment, earthquake prediction research, seismological studies, seismic design and engineering research, and preparedness planning and hazard awareness. (For a summary of NEHRP's accomplishments through 1984, see table 7.)

The national effort to prepare for future earthquakes is not, however, confined to the federal government. Around the nation, for example, scientists are constructing buildings in laboratories and knocking them down. In some very laborious efforts to learn how to design earthquake-resistant structures, researchers (primarily in the United States, but also in Japan) are creating scale models of buildings and literally shaking the slabs they rest on to see how they would fare in earthquake conditions. Some researchers build models of structures that have collapsed during earthquakes in order to see where the structural problems exist and how they can be corrected. Others concentrate on building new models to see how their revolutionary designs will stand up to any future shaking.

Still other researchers concentrate their efforts on developing new systems for *base isolation,* a concept currently viewed as having great potential in the design of structures that will resist the damaging effects of earthquakes. In fact, base isolation was the liveliest topic at the Eighth World Congress on Earthquake Engineering in San Francisco in 1984. (The Congress, held every four years, brought together fifteen hundred engineers from forty-three nations.) Base isolation is a method of isolating a building from the ground so as to decrease the impact of the ground-shaking on the building. One of the more common methods involves setting the building's pilings on giant steel-and-rubber laminated pads that absorb the vibration and thus protect the building from the violent shock waves. Another method of base isolation involves setting the building on a series of pistons containing fluid (not unlike an automobile's shock absorber system that helps prevent shocking jolts on bumpy roads). Still another method involves setting clusters of ball bear-

Table 7. Activities of Federal Government Agencies as a Result of the National Earthquake Hazards Reduction Program

Primary Agency Activity

Federal Emergency Management Agency

Arranges international and national conferences on earthquake engineering, preparedness, and seismic policy.

Supports efforts for seismic construction standards and state/local hazard reduction activities.

Develops earthquake education and awareness programs for regional use.

Provides financial assistance to regions struck by earthquakes.

United States Geological Survey

Monitors regional fault activity and prepares regional earthquake potential studies.

Engages in earthquake prediction research.

Provides data and information to the public on earthquake occurrences.

Provides seismic data to the engineering community.

Sponsors regional symposia and workshops on seismic hazards.

National Science Foundation

Studies the implications of plate tectonics for future seismic disturbances.

Studies the earthquake process in greater depth.

Conducts earthquake engineering (geotechnical, structures, and lifelines) research.

Conducts societal response (mitigation, preparedness, response, and recovery) research.

Provides support for related research both in the U.S. and around the world.

National Bureau of Standards

Engages in earthquake structural research.

Improves seismic provisions for building design codes.

Provides technical support to the building community.

Engages in cooperative worldwide structural research.

Secondary Agency Activity

Bureau of Reclamation

Studies the effects of earthquakes on dams.

continued on next page

Table 7—*Continued*

Secondary Agency Activity

Department of Defense
Develops and implements seismic construction requirements for new and existing military establishments.

Department of Energy
Develops and implements seismic construction requirements for new and existing DOE structures.

Department of Housing and Urban Development
Develops and implements seismic construction requirements for new and existing HUD multistory residential buildings.

Department of Transportation
Studies its role in assisting with transportation following major earthquakes.

National Aeronautics and Space Administration
Provides seismic data resulting from space studies (satellite lasers, etc.).

National Oceanic and Atmospheric Administration
Provides tsunami and earthquake data to the public, the scientific community, and other governmental agencies.

Nuclear Regulatory Commission
Delineates, assesses, and mitigates earthquake hazards as they pertain to nuclear power plants and waste disposal sites.

Tennessee Valley Authority
Assesses the impact of earthquakes on its power distribution systems.

Veterans Administration
Retrofits its hospitals to be earthquake-resistant.

Sources: House Committee on Science and Technology, Subcommittee on Investigation and Oversight and Subcommittee on Science, Research, and Technology, *Earthquakes in the Eastern United States,* 98th Cong., 2d sess., 1984, H. Doc. 141; Federal Emergency Management Association, *National Earthquake Hazards Reduction Program: Fiscal Year 1984 Activities* (Washington: GPO, 1985).

ings under the building's columns and walls, allowing the structure to literally roll through an earthquake and remain intact. Some methods are more experimental than others, but through

1984, approximately forty buildings, bridges, and nuclear power plants, including many in California, had been built using some type of base isolation system.[1]

The California Effort

For better or for worse, California has long been known as the progressive state, where new ideas are tried before the rest of the world even hears about them. During the 1970s, however, California was definitely not progressive when it came to earthquake preparedness and mitigation. California was the state where a motel was built on top of a major fault and was subsequently turned into a retirement home. California was the state where fewer than 10 percent of the population had bothered to purchase earthquake insurance, even though a number of companies actively advertised the availability of policies. California was the state where a school board was sued for trying to close a school they felt was too close to a fault, and where citizens consistently refused to pass bond issues to increase seismic safety in their schools. California was the state where a former large-city mayor, in rejecting seismic building codes, stated, "I would rather have a beautiful city than be completely safe from earthquakes."[2]

By 1980, however, the state began to become painfully aware of the threat of devastation. A widely circulated National Security Council report suggested that a major earthquake in the Los Angeles area (which has a 50 percent chance of occurring by the year 2015) would leave ten thousand to fifteen thousand people dead and another fifty-five thousand injured and would cause close to $20 billion in damage, crippling the region that produces a full 7 percent of the country's gross national product.[3] Further north, in the San Francisco area's Silicon Valley, the high-tech industry began heeding the warnings of seismic researchers as the companies began to realize the high stakes involved and the fragility of their existence. Computers and computer manufacturing would be temporarily shut down—just long enough for mass confusion to reign, commerce to grind to a standstill, and Japan to pick up the slack.

More by public officials' personal concern than by an outcry from the citizenry for action, the former began waging some very successful proactive efforts to prepare themselves. The state of California adopted a very stringent building code that experts point to as landmark. San Francisco is now being dotted with a variety of new earthquake-resistant buildings, relying on revolutionary designs (pyramidal-shaped structures, for instance) and base isolation systems. Los Angeles, after a long political battle, now requires a certain amount of retrofitting on unreinforced masonry buildings in the city to help them withstand earthquake vibration. Annually, the state holds an Earthquake Preparedness Week, during which time the government and an all-out media blitz instruct citizens on what they can personally do to prepare and how to do it.

Probably the most notable effort, however, has been the creation, tenacious effort, and subsequent success of the Southern California Earthquake Preparedness Project (SCEPP). When the 1980 National Security Council report was released, the state of California and FEMA launched a cooperative effort to help prepare the people, local governments, and businesses of southern California for the expected cataclysm. As Paul Flores, SCEPP's director later admitted, "We realized we were woefully unprepared for what was predicted."[4] Initially, SCEPP was launched with a volunteer interim policy board composed of interested professionals representing engineers, seismologists, school districts, the Red Cross, county and city governments, and private businesses who volunteered their time and resources to get the movement under way. With funding coming from the state and from FEMA, SCEPP was soon able to hire a staff and set up formal goals and projects aimed at earthquake preparedness, mitigation, response, and recovery. To date, SCEPP's major success has been the development of four such prototypical plans for the region: (1) a county government preparedness model, (2) a large-city (Los Angeles) preparedness model, (3) a small- to medium-sized-city preparedness model, and (4) a corporate (business) preparedness model. Each model has been developed with the cooperation of one of the entities involved. The next step, according to Flores, is to gain the commitment of all the counties, cities, and businesses in the region to adopt and implement the plans. (For a detailed discussion of how SCEPP operates and how its methods might be used in the Midwest, see chapter 7.)

Northern California businesses have also taken steps to address the seismic hazard. Silicon Valley companies, most notably Hewlett-Packard and IBM, have begun cooperative research with Stanford University researchers to mitigate the effects of a major earthquake on businesses in the region, especially as those effects relate to computer manufacturing and communications. One project involves independent room isolation systems (similar to base isolation systems) for data processing and other computer centers so they can safely ride out shock waves. With such a system, critical areas of existing buildings that were built before base isolation systems were available can be protected.

On a more entrepreneurial level, a number of small companies have sprung up offering earthquake consulting services to homeowners and businesses. Such firms, among other services, offer individual home or business inspections to evaluate hazard reduction (such as the use of bracing and gas-line safety), instructional seminars, and emergency preparedness kits (including such useful items as water containers and gas-line shut-off wrenches). Business, the firms report, is booming, a sign that California is indeed waking up to the reality of the situation. Ten years ago, such services would have gone begging.

The Mississippi Valley is, unfortunately, still sleeping peacefully save for a few voices, hoarse after years of speaking to essentially deaf ears.

Preparedness Efforts in the Mississippi Valley

Federal and Regional Efforts

One of the first entities to recognize and attempt to deal with the problems of a major earthquake in the Mississippi Valley was the federal government. It has been active on four major fronts:

Geological Research. The USGS continues to study the Mississippi Valley seismic zones and add, a painstaking bit at a time, to the current knowledge about the zones and their potential for destruction. The agency also funds regional seismological research and regularly publishes collections of professional papers on the region in a continuing effort to assemble the results of the diverse projects in a single location.

Congressional Hearings. In 1984, the U.S. Senate held a sequence of hearings on earthquake hazards east of the Rocky Mountains, concentrating on the Mississippi Valley in general and the New Madrid Seismic Zone in particular. The hearings drew upon the testimony of experts from across the nation, primarily from involved U.S. government agencies and from the various universities conducting research on the region. The hearings were designed to assess the federal government's continuing role in mitigating earthquake-related problems, especially as they relate to funding for the various agency and regional/local activity. Preliminary results of the hearings seem to indicate continued federal support and concern.

Regional Workshops. FEMA and the USGS have been cooperating for the past five years to develop regional workshops, symposia, and conferences to further the collection and dissemination of knowledge related to midwestern seismicity. One such workshop was a three-day event held in St. Louis (in May 1982), which attracted approximately seventy-five scientists and regional government officials. Another was held in Cape Girardeau, Missouri (in April 1984), and attracted over two hundred participants.

The Central United States Earthquake Preparedness Project. Realizing that it can only do so much in mitigating a disaster of regional proportions, the federal government launched CUSEPP in 1981 as a way of transferring responsibility for the problem to the Mississippi Valley region itself, where specific and concrete plans and projects could take place. In 1983, CUSEPP, under the direction of FEMA, provided funding to the governments of the seven states determined to be the most susceptible to a major earthquake in the region: Arkansas, Illinois, Indiana, Kentucky, Mississippi, Missouri, and Tennessee. The goals of CUSEPP are to (1) initiate individual state five-year plans for earthquake hazard reduction efforts, (2) establish state seismic advisory boards, and (3) participate in the formation of a multistate consortium to deal with the earthquake problem.

To supplement the effort, FEMA sponsored three major reports on Midwestern seismicity and its implications for the region:

1. *Evaluation of Past Studies and Identification of Needed Studies of the Effects of Major Earthquakes Occurring in the New Madrid Fault Zone,* by Otto W. Nuttli (1981);

2. *Estimation of Earthquake Effects Associated with a Great Earthquake in the New Madrid Seismic Zone,* by Margaret G. Hopper, S. T. Algermissen, and Ernest E. Dobrovolny (1983); and

3. *An Assessment of Damage and Casualties for Six Cities in the Central United States Resulting from Earthquakes in the New Madrid Seismic Zone,* by Allen and Hoshall, Inc. (1985).

The findings of these reports will be discussed beginning in chapter 6.

Realizing that the earthquake problem in the Midwest was a multistate issue, the directors of emergency services from the seven states involved in CUSEPP formed the Central United States Earthquake Consortium (CUSEC) in October 1983 in an effort to coodinate preparedness and mitigation activities on the regional level. CUSEC received funding from FEMA in April 1984 to carry on its efforts to develop a five-year plan. CUSEC has held a regional workshop in St. Louis (in June 1984), with future plans that include:

• Developing an interstate emergency compact in which the seven states would pledge to support one another with short-term earthquake disaster assistance (by sharing equipment, supplies, facilities, and personnel in the fields of medicine, security, law enforcement, and firefighting). This will involve assuring some form of interstate protection to those who render professional protection outside their jurisdictions (allowing doctors to practice emergency medicine in other states, for instance).

• Creating a comprehensive list of all the resources in the seven-state region that might be needed during a major earthquake (as outlined above).

• Increasing public awareness of the potential for disaster in the region through educational materials and programs, the formation of an earthquake advisory committee, and the staging of emergency "mock disaster" exercises.

On a more concrete level, the Southern Illinois University School of Medicine (Springfield) is actively planning for medical rescue efforts should they be necessary following an earthquake in the region. "Other than the Veterans Administration hospitals, there are no other earthquake-resistant hospitals in the region," warns Dr. Richard Moy, dean of the SIU School of Medicine. "As we found out so tragically in Mexico City recently, hospitals can become a net producer of casualties and incapacitate many of the physicians." Time, he stresses, is of the essence following a disaster. Dr. Moy's initial plan involves the air transport of emergency medical teams from Springfield to assist local medical personnel as needed. Dr. Moy is currently developing a model and action plan to identify available resources (such as emergency services and disaster agency resources and National Guard helicopters) and then plans to set a date for a simulation exercise. Hospitals in the Springfield area are very supportive of the plan.[5]

Dr. Moy is also working on gaining the support of other university medical hospitals in the seven-state region. Ideally, a cooperative network could be set up that would allow emergency teams of medical personnel from anywhere in the region not devastated by the earthquake to mobilize quickly and fly to the hardest-hit region(s) within six hours, assist the seriously injured, and fly those who need special treatment out of the area to hospitals not already swamped with victims. These teams could provide emergency medical care until the National Guard's medical units and federal assistance arrive (approximately twelve hours later and seventy-two hours later, respectively). Planning to seek interstate treaties to allow medical personnel to practice medicine outside their respective states, Dr. Moy stresses that "without these types of agreements, I can assure you some lawyer will sue a doctor for practicing medicine in another state without a license."

State and Local Efforts

With the federal government, and to a lesser extent regional groups, taking the lead with area earthquake preparedness, the seven states most likely to be affected by a major earthquake have done little individually in the way of preparation or public education. The states have disaster plans, but not all such plans are up-to-date and few have been tested in mock drills. Some states, like Illinois, have passed legislation that directs the states' emergency services and disaster agencies to distribute earthquake preparedness literature to schools, local governments, community groups, and media operations. In Kentucky, the Governor's Earthquake Hazards and Safety Technical Advisory Panel recently completed a major report on earthquake-related action that state needed to take in the future (information in this report will be discussed in chapter 6). Kentucky also has a building seismic code.

State geological agencies and universities are continuing to learn more about seismic activity in the region. The Illinois State Geological Survey, for example, is developing plans to drill a thirty-thousand-foot (approximately 5.7-mile) Superdeep Drillhole in the southeastern part of the state near the juncture of a number of seismic zones and faults. While the hole is designed to provide answers to a number of other geological questions, one of the goals is to better understand regional earthquake activity by studying rock formations and by placing a seismometer in the drillhole to record seismic activity. "The first twenty-thousand feet or so is sediment," explains J. James Eidel, principal geologist and head of the Mineral Resources Group of the survey. "The last ten thousand feet or so is basement rock, which we have little idea as to the composition." Ultimately, the estimated $33 million needed to drill the hole will have to come from Congress. Eidel estimates drilling will start in either the late 1980s or the early 1990s.[6]

Cities and counties, in most instances, are even less prepared for a major earthquake than the states are. There are some exceptions, though. The city of St. Louis and St. Louis County are currently cooperating on a venture to implement seismic building codes in their jurisdictions. The effort demands cooperation, however, because if one jurisdiction implements the codes before the other, it stands to lose new construction, which might locate in the other jurisdiction for financial reasons. (However, according to Dr. Otto Nuttli, a study conducted for the Building Seismic Safety Council of Washington, D.C., indicated that the additional cost for seismic safety in high-rise structures in St. Louis, if undertaken at the design stage, would be less than 1 percent.) Already, though, some large structures in the city of St. Louis are being built to specifications that are stricter than the current codes require (Metropolitan Square and One Bell Center, for example). The city of Memphis and Shelby County have appointed a seismic code provisions committee to make recommendations for a seismic building code for both jurisdictions, and prospects look good for getting seismic provisions adopted.

Another progressive city in the region is Carbondale, Illinois, primarily due to the concern of current and past mayors, a concerned city council, and a well-prepared city emergency services and disaster agency, which has been compiling earthquake literature since the early 1970s and distributing it to interested parties. Carbondale may be the only city in the seven-state region to sponsor an annual Earthquake Preparedness Week, similar to California's (though Paragould, a town in northeast Arkansas, has also taken a solid step toward informing its citizens of the importance of earthquake preparedness—in October 1984, it held its first Earthquake Festival in an effort to promote earthquake awareness). Carbondale, furthermore, has a strict seismic building code (modeled after the National Building Code) and a code enforcement officer with training in inspection, one of the few small- to medium-sized cities in the region with such a code and enforcement policy.

While a few other cities and counties in the seven-state region have seismic building codes, few have enforcement officers, fewer have enforcement officers with seismic code training, and even fewer have trained enforcement officers who conduct inspections and note discrepancies. Seismic code construction in the seven-state region is, with a few notable exceptions (such as construction of buildings using federal funds and buildings covered by Kentucky's seismic code), almost nonexistent.

A few agencies and organizations in the region (county and city emergency services and disaster agencies and universities, for example) have sponsored and/or presented earthquake workshops over the past few years. Such efforts have been helpful in drawing attention to the problem for those who attend—primarily representatives from police departments, fire departments, ambulance companies, hospitals, and other emergency-related organizations. Few members of the general public, however, attend.

The region's news media have, on occasion, gone a bit overboard—such as the newspaper that reported a very minor tremor in the area with a front-page headline reading: "Earthquake! New Madrid Fault Awakens"—but most experts in the region, from Memphis to St. Louis and all points in between, agree that the media have done an exceptional job of reporting information on midwestern seismicity, thus educating the public, keeping the topic fresh in the public's minds, and, possibly most importantly, reminding local officials of their obligations to begin serious preparedness and mitigation efforts. It is quite possible that the media may have been, and may continue to be, the best resource this region has to build the type of awareness that came so painfully slowly in California.

Why Hasn't More Been Done?

The Absence of Personal Preparedness

It is quite clear that few people in the region have taken any measures whatsoever in anticipation of a major earthquake in the region. Probably the only thing that more than a small handful of individuals have done is take out earthquake insurance. (Earthquake insurance is discussed in chapter 6.)

Even most California residents have done little to prepare for earthquakes. The state and local governments—yes; individuals—no. For example, a recent state law requiring real estate agents to disclose seismic hazards to potential buyers—more specifically, to tell them if the property they are considering is within one-eighth of a mile of a fault—has had little if any effect on property purchasing practices.[7] People are still buying property near, even on top of, faults in California. "Whole subdivisions are built on faults," adds former Carbondale mayor Hans Fischer, architect and president of Fischer-Stein Associates, Inc. (Carbondale). "When part of the fault moves, a city crew simply comes in and pours more asphalt over the crack or hump."[8]

Almost every well-known proponent of earthquake preparedness and mitigation efforts laments the general public's seeming lack of concern and obvious lack of action. Before the problem can be addressed, though, the sources of the problem need to be addressed. There are at least the following reasons for such lack of concern:

Lack of Information.   While many people assume everyone has heard about the coming cataclysm, this simply is not the case. Not everyone reads newspapers, watches television news, or listens to the radio. A recent poll conducted by a regional newspaper found that there were people in the area who had not even heard about the possibility of a damaging earthquake in the near future.[9]

Inadequate Information.   Even if people are aware of the problem, there is no guarantee they will act. Sociologists and psychologists have found in numerous studies that knowledge of a problem doesn't necessarily lead to action to solve the problem. "Merely being aware that a destructive magnitude earthquake could occur in one's region of the country is not sufficient to motivate either individuals or communities to take preparedness or mitigation actions," states Dr. Joanne M. Nigg, assistant professor in the Center for Public Affairs and acting director of the Office of Hazard Studies at Arizona State University (Tempe).[10] One of the reasons people do not take action is that they do not know what action to take. While there has been a great deal of publicity about midwestern seismicity, there has been almost no publicity about what individuals can do to prepare for it. "We conducted a survey in 1984 to see what people knew about earthquakes in the region," says Lawrence Malinconico, assistant professor of geophysics at Southern Illinois University at Carbondale. "The results were very discouraging. People actually knew very little, and much of what they thought they did know was inaccurate." He reports that a school teacher who asked him to speak to her class said that if an earthquake occurred, she would have her students sit against the brick wall just as they would for a tornado. "In a major earthquake, being against a wall is about the worst place to be," warns Malinconico. "It's going to fall on them."[11]

Other Priorities. Since damaging earthquakes are not an everyday occurrence in the region, preparing for one simply is not a priority among the general public. There are enough pressing problems to contend with each day without adding one more to the list. "Even if you live in 'Tornado Alley' or along the San Andreas Fault, you don't think about tornadoes or earthquakes too much," explains Catholic University's Frederick Ahearn. "You don't get up every morning and wonder if there's going to be a disaster that day. There are too many other things to deal with."[12] An insurance corporation executive notes that during the 1970s, only about 8 percent of California homeowners purchased earthquake insurance. A number of companies decided to advertise the policies, but to no avail. After a moderate earthquake, however, the companies were bombarded with requests. The following year when the premium renewal notices were sent out most of those who had purchased the insurance asked that the earthquake rider be taken off.[13] The same situation occurred in the Midwest just a few years ago. Following a moderate tremor, the phones at one of an insurance company's regional offices rang off the hook for a week with people wanting earthquake insurance. Earthquakes, for a short moment, had become a priority, but only for a moment.

Denial.   There are a number of people in the region who, even though they are aware of the fact that a damaging earthquake is inevitable for the area, refuse psychologically to believe it fully. "The predictions are so far-fetched, most people here don't give it a thought," states one resident.

Lack of Experience.   If a person has not experienced something before, it can often be hard for him or her to get concerned about it happening in the future. The frequency of strong earthquakes in the Mississippi Valley is extremely low. The last relatively major one occurred in 1895. Only lifelong residents into their nineties would have even been born at that time. At least one expert noted off the record that he would like to see a mild to moderate tremor in the region soon to "wake people up" to the problem of the larger one that is predicted.

"Crying Wolf."   There is truth to the conventional wisdom that if one threatens or promises something for long enough without making good that threat or promise, people will begin to ignore it. Such may be, or may soon become, the case in the Midwest, where a number of residents say they are tired of hearing about the predictions. This is certainly the case in California, where citizens have been bombarded with predictions and admonitions for two decades.

Lack of Responsibility.   There is a segment of the population who, even though they accept the inevitability of the disaster, assume that someone else (usually the government) will take care of them afterwards and that it is someone else's responsibility (again, usually the government's) to handle any and all preparations for the disaster. They do not feel they are responsible for taking care of themselves or for engaging in preparation themselves. "We found a strong citizen expectation that it was important for government to take steps to lessen the effects of large magnitude earthquakes," states Arizona State University's Joanne Nigg.[14]

Need for Risk. There are a number of people who, for any number of reasons, need and even seek risk. Life to them is too boring unless there is some element of risk involved. Such people view the eventuality of a major earthquake in the region more with excitement than with consternation; any preparation efforts would simply diminish the anticipation and excitement.

Invincibility. Closely allied with those who enjoy risk are those individuals who feel that if and when the disaster occurs, it will not affect them directly. "The vast majority of people bet, and quite correctly, actually, that they won't be involved directly in a disaster," argues E. L. Quarantelli, professor of sociology and director of the Disaster Research Center at the University of Delaware (Newark)—and one of the world's most respected authorities on the socioeconomic effects of disasters.[15] Even if such people recognize that they might be involved in a disaster, they feel that nothing serious will happen to them, and research has shown that unless people see themselves as being physically endangered, warnings do not carry much weight.[16]

Fatalism. One of the most common responses people in the Midwest give for not taking steps to prepare for earthquakes is that if something is going to happen, it is going to happen, regardless of anything they do about it. They seem to feel that whether they prepare or not is ultimately inconsequential.

Other Fears. Some people report that they are more concerned about nuclear war and other salient issues than they are about earthquakes. They make the distinction between human-made disasters and natural disasters and claim concern only for the former. They often hold to this belief even after they are told that while nuclear war and other such disasters are only possible a major earthquake in the region is truly inevitable.

Lamenting the lack of action caused by such attitudes is fruitless. Addressing the opinions and trying to change them is the only way that those who encourage earthquake preparedness and mitigation efforts among the general population will achieve any measure of success. Even then, they can only expect to influence a small percentage of people. Most people can be expected to cling to their beliefs, regardless of the validity of those beliefs.

The Absence of Public Preparedness

There are two major reasons why local midwestern governmental bodies and other interested professionals have not engaged in earthquake preparedness and mitigation activity:

Cost.   Simply stated, it costs money to engage in just about any type of earthquake preparation or mitigation effort. It costs money to develop an earthquake preparedness plan. It then costs money to test it. It costs money to buy needed supplies or consulting services. It costs money to implement building codes. It costs money to hire and train someone to enforce those building codes.

Most important, as some cities, counties, and states see it, it costs money when businesses and contractors decide to build somewhere else. Such decisions might result if a city, county, or state has—and enforces—a stringent building code, while surrounding jurisdictions do not. (Kentucky is the only state in the region with a seismic code.) Such decisions might also result if there is too much publicity about the earthquake potential in the region. Corporations and other outside investors, it is felt by many local and even state governments, will look elsewhere if it becomes obvious to them that the Mississippi Valley is due to erupt someday.

This is the financial concern. Its validity (or lack of validity) will be discussed in chapter 6.

Lack of Pressure. There is no question that pressure groups influence governmental decisions. It is equally obvious that there are no vocal earthquake pressure groups to speak of in the Mississippi Valley. In fact, there are few such groups anywhere. E. L. Quarantelli has only been able to identify a dozen disaster pressure groups in the nation, and most of them eventually resulted in abortive attempts. In California, he has looked all over the state for emergent citizen groups and found only two that could remotely be called preparedness groups. (According to one public official in that state, a public opinion poll ranked concern for earthquakes tenth on the worry list, right behind pornography.[17]) In the Midwest specifically, "there are no constituencies that give a damn," contends one environmental engineer, "and legislators aren't going to act without them."[18] On a more local level, the problem is the same.

At least two conditions need to be present before most activist groups can get off the ground. First, there needs to be an identifiable, easily culpable, yet harmless (in terms of retaliation) enemy. For instance, government and corporations are regular targets of pressure and activist groups because they have high profiles, can be slandered with little difficulty, and tend not to take any retaliatory action against the activist groups that are attacking them. There is no such enemy as far as earthquakes are concerned. Second, the enemy needs to be doing something that the groups want stopped, rather than not doing something that the groups want started. For example, there are a number of groups protesting nuclear power, the pornography being sold in corner convenience stores, increases in utility bill costs, and South African racial policies. All of these groups are demanding that certain activities *cease.* It is more difficult to find activist groups insisting that the chosen enemy *begin* something. (For one thing, if protesters insist that something cease, they have no further responsibility. If they insist that something be started, however, someone is liable to ask them to help start it. Any group that begins demonstrating for better civil earthquake preparedness and mitigation efforts must know in advance that its members will be taken to task if they do not participate in the mundane work required to reach such goals.)

The reality of the situation, in other words, is that it is highly unlikely that the earthquake problem in the Midwest will ever become hot enough to cause any activist group to form and insist that anything be done. The activist groups will undoubtedly form *after* the earthquake and demand to know why more was not done ahead of time. Change, it seems, will have to occur through more conventional channels. Earthquake preparedness seems to be an issue that will require the government to pressure the people, instead of the other way around.

PART THREE: THE FUTURE

6. The Next Twenty Years

As discussed in chapter 5, there are major concerns about who will take responsibility for earthquake preparation efforts in the Midwest. There is, as Dennis Mileti, professor of sociology and director of the Hazards Assessment Laboratory at Colorado State University (Ft. Collins) argues, a great deal of leeway in the system for shifting the burden of responsibility: "The federal and state governments are in the best position to provide earthquake response planning, but they don't want to try to tell local governments what to do. Local governments have the responsibility to take action, but they don't have enough information. So the people who can help won't, and the people who can't help will be expected to. A local government official *may* be able to do a good job, however. If so, it will be the result of that person's personality, not the result of the way government works."[1]

There is also the previously discussed concern that taking action to mitigate disaster will simply draw unfavorable attention to the region, causing concerned citizens to leave and causing new business to shy away from the area. Research suggests this not to be the case, however. In concluding one study, for example, the authors state, "The findings should reassure local government officials and policy makers that earthquake predictions—at least those couched as more general announcements—will not generate widespread, permanent out-migration."[2]

Experience also suggests this to be true. California's population continues to increase, and business is booming. Companies continue to build in California. Business is also thriving in and around Charleston, South Carolina, which had a major earth-quake in 1886. As Hans Fischer states, "I don't think our region has lost, or will lose, any new business just because of being in a seismic zone."[3] The question, then, remains: What can be done to better prepare the region for the next powerful New Madrid earthquake?

Public Preparedness

Seismic Design

Experts agree almost unanimously that the best and most im-portant step a county or city can take to prepare for future earth-quakes is to implement and enforce seismic building codes. This one step will do more to save lives than any other step that can be taken.

A number of people argue that structures built to withstand heavy winds are also built to withstand earthquakes. To a degree, this is true. "Buildings designed for heavy wind loads provide a *degree* of earthquake resistance," according to Anshel Schiff, professor of mechanical engineering at Stanford University and formerly with the Center for Earthquake Engineering and Ground Motion Studies at Purdue University.[4] Wind loads determine the amount of steel or other building material to be used in a build-ing. The major difference between seismic design and wind load design, though, is that seismic design involves more than just absolute strength. Seismic design involves the detailing associ-ated with construction—how close, for instance, the retaining steel and reinforced concrete are to be placed. Nonseismic re-sistant structures, according to Dr. Schiff, tend to be more brittle. When they exceed their loads, they collapse. Buildings designed for earthquake resistance, on the other hand, have ductility built into the structures, which means that when the design loads are exceeded, failure and cracking may occur, but collapse and loss of life will probably not occur.

It is not difficult to create seismic building codes for local communities or counties, because adequate state and national codes already exist. As Hans Fischer notes, "There are a lot of good codes already available. They just need to be adopted and enforced." (Enforcement—the real problem after codes are adopted—requires hiring and training staff to make sure the codes are adhered to; such codes at present are not uniformly enforced, because most smaller communities have no checking capability, and even larger towns often only have a draftsperson checking blueprints, not an engineer checking seismic design.)

The cost of building to seismic codes is not exhorbitant. Some studies suggest that adaptation to seismic building codes adds 30 percent to building costs, but "that's simply not so," according to Schiff. "If you have *good* construction in the first place and add seismic provisions, you're talking only a 3 percent to 5 percent increase for most construction." (In some cases, as noted in chapter 5, those increases are even smaller—less than 1 percent.) The problem of construction projects moving elsewhere to avoid such costs, of course, is another concern. The real question, however, is a public one. "What priority does life and public safety have in economic development?" asks Fischer.

Another step that can be taken in tandem with the implemention and enforcement of building codes is the inspection of existing structures. There are, moreover, certain critical facilities, such as power plants, hospitals, and fire stations, that need special attention. If these facilities are damaged during an earthquake, the impact will be much greater than the damage itself; critical services will be jeopardized. Schools and churches should also receive special attention to assure that they are earthquake resistant.

Another concern is whether or not existing structures that are not earthquake resistant should be retrofitted (strengthened or partially rebuilt). In general, retrofitting involves providing braces for walls so they will not collapse during an earthquake. The walls are tied together so the structure will not move in all directions and end up with portions bumping against one another. Retrofitting can also involve constructing additional adjacent members (such as buttresses) to take the force of the earthquake and transmit it to the ground and away from the structure itself.

While it may be prudent to consider retrofitting critical structures, the cost of retrofitting other structures is realistically prohibitive. "I don't think retrofitting will be implemented to any extent in the Midwest," states Arch Johnston, associate professor of geological sciences at Memphis State University and director of the Tennessee Earthquake Information Center. "In Memphis," Johnston continues, "we'd like to get building codes implemented for new structures, then possibly look at a very restricted class of buildings for retrofitting, such as the old masonry schools. Across-the-board retrofitting is very unlikely, though."[5] Fischer agrees: "Cost-wise, it's almost impossible to retrofit schools and other public buildings that are at high risk." A cost analysis should be performed on such structures, though, to see if retrofitting *is* realistic.

A balanced approach to building codes and retrofitting was probably summarized best by an editorial in the *Southern Illinoisan* (Carbondale) on October 6, 1985: "Reinforcing every building in Southern Illinois is not the issue; the cost would be unaffordable. But examining and strengthening senior citizens highrises, hospitals, schools and other large, potentially vulnerable buildings ought to be a priority. And any new construction, especially of large or public facilities, is equally important. Every year of new construction that replaces unsafe buildings is another layer of protection for people and property."

Lifeline Management

During a catastrophe, management of gas and electric lines and water and sewer mains is very important. If lifelines are damaged, not only are critical services disrupted, but secondary problems (such as fire and disease) can result. Gas lines that break can lead to raging fires, as can downed electrical power lines. If water mains break, the fires will go unattended, since there will be no pressure for the hoses. If water and sewer lines are damaged, residents will be without fresh water, and disease can spread.

Power substations need to maintain an adequate supply of replacement parts, the lack of which will be the main problem after an earthquake. (The power lines themselves will generally survive the tremors.) The porcelain members on the equipment at the substations tend to break easily, and if there are not enough spares on hand to replace the broken ones, power transmission will come to a standstill. In addition, base isolation systems should be used when building these substations whenever possible, so as to limit the shaking the equipment is subjected to.

Broken water and sewer lines are common problems in communities, especially those where the lines are old and brittle. City crews can spend much of their time doing nothing but repairing breaks. In regions not prone to earthquakes, this practice may be cost effective: it may cost less to constantly repair breaks than to completely replace the system until such a time as the breaks are occurring too often. In an earthquake-prone region, however, this practice may not be particularly wise. Completely replacing old, brittle systems may be worth considering, since such systems will invariably collapse entirely during a strong earthquake and thus completely cut off the community's fire-fighting ability and drinking water supplies. "Currently, however, the replacement cost is tremendous," admits Schiff, "so no communities are really doing it."

Only when the cost of repairing the leaks becomes so great that simple economics justifies replacing the whole system do communities consider seismic water and sewer lines. Such earthquake-resistant water pipes incorporate flexibility and ductility. Flexibility is most needed at connections between the pipelines and larger, more rigid structures (such as tie-ins at pump stations, valve structures, storage tanks, and bridge abutments). In terms of ductility, pipelines composed of continuous ductile steel and protected against corrosion are among the strongest and most resilient structures during an earthquake, especially when sharp changes in elevation and direction are not present and when the pipes are buried at a relatively shallow depth. Having such a modern system can mean the difference between a community's survival or collapse during a major earthquake.

Regional Planning

As discussed earlier, earthquake preparedness and mitigation efforts must originate with public officials and filter down to the general public. While desirable, it is simply not realistic to expect the general public to put any pressure on government officials or anyone else to begin preparation or mitigation efforts. The key to success is to get emergency response groups to gear up.

Communities and counties must engage in direct predisaster planning. For one thing, during a major earthquake communities will be effectively cut off from one another and from most other forms of outside assistance. They will simply have to take care of themselves—to survive on their own. While society has certainly progressed since the early nineteenth century, today's communities will be in little better shape following a major earthquake than were New Madrid and Little Prairie in 1811, because the conveniences they rely on today (such as highways, utility systems, and telecommunications) will be rendered useless.

In terms of emergency communications, for example, one expert warns that, "in general, emergency response departments at the various levels are very ill prepared. One that I visited had an emergency response center and radio equipment, but the radio equipment was sitting on a card table. And this was one of the better-prepared centers I visited." He reports on another emergency service organization that purchased the same communication system as the state police, but failed to reserve a channel so they could get into the network. Emergency communication systems that rely on telephone lines will be useless, because many lines will be down and everyone will be on the remaining lines trying to call relatives. In most disasters, emergency personnel usually end up talking to each other through the windows of their cars. "They're simply not prepared with the right kind of emergency communication equipment," adds the expert.[6] It is imperative, then, that city and county emergency service organizations plan how they will communicate. (Walkie-talkies, for instance, will be absolutely essential.)

An excellent model of a coordinated, overall emergency plan to study and learn from is that of the Southern California Earthquake Preparedness Project (SCEPP), which was briefly mentioned in chapter 5. SCEPP's effectiveness has been due to its ability to help local entities (counties, communities, and businesses) take responsibility for their earthquake preparedness activities and to encourage them to begin developing and implementing such plans. SCEPP helped to develop prototype models for (1) counties, (2) large cities, (3) small- to medium-sized cities, and (4) corporations and other businesses. The first two years were spent designing these "planning partnerships." SCEPP arranged with one example from each of the four categories to help it develop specific plans tailored to its particular needs. For example, SCEPP worked with one county's board in the region to help it decide what it would be responsible for and what type of preparedness program it would want for its county. Once this county's plan was developed, it became the model for other counties to adopt. The same procedure was used to develop large-city plans, small- to medium-sized-city plans, and corporate plans. (For an outline of SCEPP's prototype plans, see appendix 2.)

The distribution of model plans occurred through a series of county conferences. "We received the commitment of each of the county board of supervisors and invited the 'stakeholders' in the area to attend," explains SCEPP's Paul Flores.[7] These "stakeholders" included governmental officials, engineers, university earth scientists, business and industry leaders, school officials, and heads of volunteer organizations. (The sessions were not open to the general public.) SCEPP provided these leaders with planning information and explained how to disseminate it to their constituencies. The leaders were impressed with the presentation and the fact that SCEPP was distributing this information and literature to them free of charge. (It was the first time such a cross-section of professionals had the opportunity to meet and talk together.) Following the main sessions,

there were specialized workshops that SCEPP had developed—
one for county officials, one for city officials, one for school
officials, and so forth. SCEPP tried to impress on everyone that
it (SCEPP) was not ultimately responsible for implementing the
plans. Those directly involved were. "We also impressed on them
that when an earthquake occurred, it would be *their* loss if they
hadn't prepared, not SCEPP's loss," adds Flores. SCEPP is currently in the process of following up with the counties to see
how their individual plans are progressing.

Flores sees the danger of institutionalizing SCEPP. "It is vitally
important that the local entities take charge of the effort," he
emphasizes. SCEPP's push is to continue to help these entities
realize their responsibilities. Earthquake preparedness should
be institutionalized at the local level, he feels, not at the state
or regional level.

SCEPP's experience holds significant value for Mississippi Valley preparedness efforts. In fact, the Federal Emergency Management Agency is currently developing a system to make SCEPP's
material available to planning groups in the Midwest. "If I had
one piece of advice to give to your regional planning group,"
states Flores, referring to the Midwest, "it would be to develop
such models with one of each of the four entities and then offer
these model plans to *all* of the counties, communities, and businesses in the region. You have to start somewhere. Develop a
solid link with one county, one city, and one corporation. Then
begin to decentralize the program once these models have been
completed. Once you have a prototype program for each entity,
it is easy to transfer it to the other entities. The program will
have their respect and will thus be more readily accepted."

Medical Preparedness

As discussed earlier, few, if any hospitals, except for the Veterans
Administration hospitals, are resistant to earthquakes. As such,
they are liable to be net producers of injuries and deaths during
an earthquake instead of sources of assistance for the injured.
In the 1971 San Fernando, California, earthquake, for instance,

four hospitals collapsed. It would behoove every community in the Midwest to consider some seismic retrofitting of their hospitals. (Most hospitals in the region are ill suited for disasters in other ways, too: most do not have emergency power sources, except for lights; most do not have emergency water supplies.) Hospitals and other health care facilities, for example, should schedule and execute regular earthquake drills, but these may only be effective if the buildings themselves are earthquake resistant. If they are not, the medical personnel and other staff will have no time to care for outside patients following a major earthquake. They will have to take care of themselves and may even, as was the case in Mexico City, have to rely on outside help.

It may be worthwhile, furthermore, to consider planning for what, under other conditions, might seem like drastic measures. During the 1964 Alaskan earthquake, for example, two of the four hospitals in Anchorage were destroyed. One of the two remaining hospitals was a government hospital, and it was so bound by regulations that by the time it found out what it could and could not do, it was too late. As a result, the entire burden fell on the one remaining hospital. That hospital did amazingly well because the chief of staff simply took charge of the situation, telling all the other doctors that, for the time being, he was the chief resident and they were all interns. He then began giving orders to the doctors, assigning them to certain places and shifts. "Such a method might be required here if medical assistance is to go smoothly and effectively," states Dr. Richard Moy of SIU's School of Medicine.[8]

While the physical well-being of victims of the disaster will be of paramount importance, it is also important to make plans for the mental well-being of many of the victims, who may be in shock or experience other psychological trauma. Agencies providing this type of service should consider developing postdisaster plans for such victims, especially counselor training in both the short-term and long-term effects of such disasters.[9]

It is imperative that the medical professionals, agencies, schools, associations, and services as a whole in the seven-state region begin and continue plans to work together. In that regard, Dr. Moy's grassroots planning efforts deserve special attention and consideration. The first step in any such effort is to identify what resources are in place and what will be needed. Since it is reasonable to assume hospitals will be out of commission, for example, "we would like to set up emergency response centers at local airports—bubble buildings that could be kept in storage and quickly blown up when needed to serve as medical centers," Dr. Moy explains. When the medical teams arrive in helicopters, they will have places already set up so they can begin work immediately. (One of the problems, of course, is determining a way to get the injured from the hospitals, where they will naturally be congregating, to the airports. This, he admits, still needs to be worked out.) Those who need special treatment can be air evacuated to hospitals outside of the area of destruction.

Once a prototype model is set up through the SIU School of Medicine in Springfield, Dr. Moy hopes to gain the cooperation of other university medical schools in the region, offer them the model, and arrange cooperative agreements. Again, as mentioned in chapter 5, this cooperation will require interstate agreements that will allow medical personnel to perform their services in an emergency situation in another state without the threat of lawsuits. But if the dream becomes a reality—and it is reasonable to assume that it will—the Midwest will be blessed with trained teams of medical personnel ready to fly at a moment's notice to any area of damage and destruction in the Mississippi Valley and begin medical relief work immediately.

Public Awareness

The people of Memphis are much more concerned about the earthquake problem than they were four or five years ago. According to Memphis State's Arch Johnston, "We've had a quantum increase in the number of requests for talks and tours of our Earthquake Information Center." Public education and information dissemination, in other words, while a formidable task, can be effective.

In addition to being effective, it is necessary. "In the Midwest, the idea is to get the people to talk about the risk," states Frederick Ahearn, of Catholic University. People will keep their worries about the risk at the subconscious level and deny the problem unless they keep being reminded, he feels.[10] As pointed out in chapter 5, furthermore, much of the information that people do have is inaccurate and must be corrected. "One problem we have is that the public seems to think we know when a major earthquake is going to occur and will tell them," cautions Lawrence Malinconico. "Of course, that's not true, and it's important that we get the message to them that we don't know when to expect it."[11]

The city of Carbondale, Illinois, has taken the lead in generating public awareness with its Earthquake Preparedness Week, but, frankly, interest has been weak, except among professionals and the media. Widening the effort to a regional event might generate more interest, certainly among professionals, and possibly even among the public in general. Something like this would have to be coordinated through the Central United States Earthquake Consortium in Marion, Illinois. Naturally, the media would play the pivotal role in publicizing the event and maintaining awareness throughout the week with challenging stories in their newspapers and on their news reports and concrete tips on what the public can do to prepare for the earthquake.

The best time, however, to get information to people effectively is right after another major earthquake has occurred somewhere, according to E. L. Quarantelli of the University of Delaware. He recommends that people hear about New Madrid immediately after a major earthquake in, for example, Mexico or Japan. This strategy works especially well when trying to impress upon emergency organizations and local governments the need to prepare.[12]

Then there is the concern about what information, specifically, needs to be disseminated. No one has yet been able to establish a link between earthquake education and people taking appropriate action, according to Dennis Mileti ("You can put together the best public education campaign in the world, and people will become more educated, but that won't necessarily mean they will become safer"), but an effective education campaign must, he contends, do three things:

1. Provide people with information about the risk. This includes scientific information about the causes of earthquakes, the history of their occurrence, the state of prediction techniques, the estimates of future damages, and so on.

2. Tell people where they can get more information. "If you're lucky enough to spark their interest, they'll seek more information, so tell them where they can get it," Mileti suggests.

3. Tell people what to do. If people are not told exactly what to do to prepare for an earthquake, Mileti contends it should come as no surprise if they do something else.

Essentially, there seem to be two excellent avenues for dispensing information about earthquake preparedness and mitigation in the Mississippi Valley. The first is the school systems, especially at the elementary and secondary levels. "We need to start teaching the children," suggests Malinconico. "It may be too late to teach the adults." Kentucky is already taking the lead in this area. Schools there are taking steps to implement earthquake preparedness lessons in their social science and general science classes.[13]

The second avenue is the media. As mentioned in chapter 5, the newspapers and television and radio stations have been doing an excellent job of keeping the issue in front of the public. All regional experts have pointed to the good work they have done, and they and all the national experts recommend more of the same. The media can do an excellent job of educating people to the risk their community is in and giving them instructions on what to do, suggests Ahearn. (Brochures and other public information campaigns, on the other hand, are probably not very effective, according to Paul Flores: "Frankly, I'm very pessimistic about them. They just don't seem to work. The best way to continue to get information across to the public is the media.") "I think a year-round effort by the media would be very helpful," adds one source who wished not to be identified. "This effort, however, should address public apathy and get into some controversial areas. For example, the media should play a greater role in promoting the idea of reducing hazards, not just in making people conscious of the problem. They should begin to play 'devil's advocate' in order to get changes to actually occur at the local level."

A potential problem with the media, though, is the time element. "If a major earthquake doesn't happen in the Midwest soon, I would expect the media to give up on it," states Stanford's Schiff. "It may be hard to keep their attention until one occurs."

Business and Industry Preparedness

Business owners and industry leaders must take measures to protect their assets and people. It is important that businesses and industries protect their assets in order to be able to survive a major earthquake and to be able to resume business quickly in order to maintain their market shares. These same businesses and industries, of course, have a moral (and legal) obligation to protect their employees while they are on the job and their customers while they are on the premises.

Companies in the region should determine the risk of their not retrofitting their businesses and plants and make appropriate decisions. These decisions, again, will be based on such factors as employee safety and the potential for disruption of business, damage to assets and loss of market share. In general, to be earthquake resistant, buildings should be made of reinforced concrete (or wood, if appropriate), be solidly braced, and be absent of wide, unsupported spans.

Company disaster plans should be complete, effective, and updated. Drills should be held occasionally so glitches can be worked out and so employees are aware of what to do. Disaster plans differ by type of disaster, so a general-use disaster plan is inappropriate. Tornado disaster plans, for instance, recommend standing up against a wall. Fire disaster plans recommend evacuation. Both actions are extremely dangerous if taken during a powerful earthquake.

Business owners should ask their insurance carriers to act as consultants in devising appropriate earthquake disaster and recovery plans. Most reputable business insurance carriers have such resources available, either in written form or in staff expertise—or preferably in both. When an insurance carrier is not able to provide a business with the kind of information and resources it needs to engage in effective earthquake preparation and mitigation, it might be wise for that business to find a carrier who does.

Computer and data processing centers or departments require special care. While there is no question that human life is more valuable than computer records, there is a much greater chance computer activity will be disrupted during a moderate earthquake than there is of employees being seriously injured. Thus, while a business's first priority will be to ensure the safety of its employees, the second might be to ensure the safety and continued functioning of its data processing center or department.

In cases where a whole building is not earthquake resistant, there are two recommended techniques to protect the computer and data processing centers. One is to have a separate earthquake-resistant structure that computer operations can be moved to after an earthquake if the permanent center is damaged. While this is the less expensive of the two alternatives, it is the less effective one, too, because operations are bound to be disrupted for a period of time after the earthquake. Thus, even though a business may have the capabilities to start again, valuable information can be destroyed in the meantime. The second option is being tried by many companies in California (as described in chapter 5). These data processing centers are being built with base isolation systems, even though the buildings that house them are not earthquake resistant. The most sensitive and critical parts of the structures (the data processing centers) are, therefore, protected. This latter option is not inexpensive, but it might be a very wise investment if the business's data processing operations are critical to its overall operations.

Again, each company must assess its own perceived risks and balance them against costs. Part of *any* data processing center efforts, however, must be a consideration for alternative, independent sources of power, since outside power probably will not be available after a large earthquake. If, further, the center relies on electronic transmission of data, the business should expect communication lines to be down and thus have a backup system to allow critical information to be protected and transmitted manually. An insurance carrier may be of assistance here, but because the development of such backup systems are such a specialized field, it might require the services of a private consultant.

Banks are among types of businesses requiring special consideration. What would the economic impact to the area be with communication lines down and financial data (and assets) in ruin? If there is one type of business in the region that is most vulnerable to a major earthquake, it is banks. A region can survive without utilities for awhile; and food, water, and medical supplies can eventually be shipped in. But if a region is without banking, commerce is effectively shut down, and this can literally cripple a region economically, even to the point of not being able to recover. A host of other specialized businesses—chemical dump sites, malls, rail lines, coal mines, media operations, insurance companies, gas stations, and mobile home courts, for example—must also assess their value to the region and what impact their inability to function (or their contribution to the problem) would have.

In the final analysis, business owners should ask themselves: What would the problems associated with the destruction of my buildings and the disruption of operations be? If limited, the business may choose to take no action. If numerous or major, good business sense might dictate that some action be taken.

Scientific Endeavors

By and large, the government and the scientific community are doing a credible job of increasing their knowledge of midwestern seismicity. There are, however, at least two possibilities for research that might be worth pursuing.

First, while such study has not received much attention in the United States, where it is looked on as being highly speculative, it might be worthwhile to study anomalous animal behavior along known faults in the Midwest. If, as some scientists suggest, some animals *do* respond to radon, other gases, or electromagnetic activity that precedes some earthquakes, such an effort might ultimately prove valuable (though realistically, as was mentioned in chapter 4, funding for such research would be difficult to obtain, and finding reliable volunteers who would be willing to participate over the period of years required to assemble any meaningful data would be equally difficult).

Second, it might also be worthwhile to conduct further research on some of the possible correlations between Mississippi Valley earthquakes and such phenomena as moon phases, sunspots, comets, and seasonal variations. While preliminary investigation has turned up very little valuable information, further study might find more of substance.

Individual Preparedness

Financial Preparation

Earthquake insurance is currently a lively and controversial topic around the nation, especially in California, where some insurance companies have sold so many policies they are becoming concerned about their ability to pay off claims after a major earthquake, particularly if the quake occurs near Los Angeles or San Francisco. Earthquake insurance has been offered by the industry since 1916, when it began, naturally, in California. Until the 1971 San Fernando tremor, interest in such insurance was limited. Since the tremor, however, interest in California has steadily increased. In fact, over 75 percent of all earthquake insurance policies written in the United States today are written in California.[14]

As pointed out above, there is such a demand for earthquake insurance in California that some companies have begun questioning the wisdom of selling it. They are not sure they would survive financially if a major earthquake struck and they had to honor all their policies. When these companies started refusing to allow policyholders to add earthquake coverage to their homeowners' policies, though, the California state government stepped in and ruled that if an insurance company offered homeowners' insurance to a policyholder, it must also offer earthquake coverage. (The policyholder is not required to purchase the additional earthquake coverage, but he or she must at least have the option of purchasing it.)

The insurance industry has reacted to this new requirement in two ways. Some individual companies, most of them smaller ones, have simply decided to get out of the homeowners' insurance business altogether, since they did not feel they could afford to sell the additional earthquake coverage. Others, particularly the larger ones, have begun to discuss the problem within their professional associations, suggesting that the federal government should be required to back up the insurance industry financially if claims that result from a major disaster threaten the fiscal solvency of the insurance industry. Currently, several proposals are being discussed in the insurance industry on what role it would like to see the federal government play in earthquake reinsurance (the assumption of all or part of the risk undertaken by an insurance company by another insurance company or by another entity, such as the government). Under one of the proposals, the federal government would be required to begin paying off each insurance company's policyholders once that company had paid out the amount of premium it had collected over the years, thus leaving the insurance company still fiscally solvent.

In the Midwest, the situation is a little different. "Companies are writing more and more policies in the region," states one insurance company executive, "but I don't know that any of them have any idea how great their losses would be. Insurance companies are in business to write policies, but some of them are starting to take a look at how much they're writing in the Midwest. Some of them soon may say, 'No more!' Others may limit their policies to certain types of structures." These structures would undoubtedly be small wood-frame residences, which traditionally have the best survival rates during earthquakes and sustain the least damage. Unreinforced masonry structures, especially the older and larger ones, represent a high risk, since they traditionally sustain the greatest damage during earthquakes. Commercial property is also considered a high risk to earthquake insurers. Such insurance is still generally available for commercial buildings, but the more valuable the property is, the harder it may be to get earthquake insurance for it in the future, as it may be for high-rise structures in general.

The desirability of purchasing earthquake insurance depends on how much risk one wants to assume. Contrary to what many people believe, homeowners' insurance does not automatically protect against earthquake damage. Homeowners who want earthquake damage protection must ask their agents to add riders to their policies to cover earthquakes. The premiums, at least for now, are reasonable. The deductibles, however, can range from 2 percent in the Midwest to as much as 10 percent in California. The lower figure means that policyholders must pay for the first 2 percent worth of damage to their homes or buildings. In other words, if a house is worth $50,000, if the homeowner has a 2 percent deductible, and if an earthquake causes $1,000 worth of damage (2 percent of the value of the home), the homeowner must pay the $1,000 out of pocket for repairs; if the earthquake does $5,000 damage, the homeowner pays the first $1,000, and the insurance company covers the remaining $4,000.

Earthquake coverage is, of course, designed to protect from catastrophic expenses. If a small tremor causes a few cracks in a foundation, a homeowner can purchase spackling compound and repair it. If such a small amount were covered by earthquake insurance, however, the homeowner would be more likely to hire a professional to do the work. Insurance companies, in other words, are able to keep their premium rates low for earthquake insurance because they are not required to pay for large numbers of small claims, many of which could be done by the homeowners themselves. To provide coverage with a lower deductible would require a much higher premium.

"The government will take care of me" is a common reply people give when asked about their lack of insurance coverage. Following major disasters, the media always make much of the issue of federal aid: Will or will not the region be the recipient of it? Many people assume that federal aid means the government will pay for the value of what people have lost, just like an insurance company will. Actually, federal aid simply means the people in the region are eligible for low-cost loans to rebuild. In other words, if an earthquake destroys homes, the federal government is not going to pay the homeowners for losses. The best they can expect from the government is a low-interest loan to rebuild.

The question still remains as to whether or not homeowners in the Midwest should purchase earthquake insurance. The experts say a major earthquake in the region is inevitable. The experts also say that one of the best things a person can do to prepare for the earthquake is to purchase earthquake insurance. The answer to the question, then, really depends on whether or not one believes the experts. Human nature being what it is, it is not unrealistic to expect that more people will want the coverage as they begin to find out it is getting more difficult to obtain. When something is easily available, people tend to take it for granted: as long as earthquake riders are easy to obtain, demand should be steady; if and when it becomes difficult to obtain, demand will undoubtedly skyrocket. The result may be identical to what is currently happening in California, where all factions (insurance companies, homeowners, and the government) are assessing their financial risks very carefully.

Preparations Around the Home

Making a home earthquake proof is probably impossible, especially if one lives near the epicenter of the next major cataclysm. Even if a home is designed and built using every earthquake engineering technique known, it will still not survive an extremely powerful earthquake that occurs directly underneath it.

The relative risks of earthquake damages do depend, however, on how a home is built. A well-constructed wood-frame house, everything else being equal, will survive much better than a masonry (such as brick or stone) home, because the wood is more flexible and moves with the earthquake, whereas the brick or stone, being more rigid, simply cracks. In addition, a home built on a concrete slab is likely to sustain less damage than one with a basement (especially if the basement is built of unreinforced masonry blocks). The blocks will tend to crack during a major earthquake, and even a well-constructed wood-frame house can be in danger of partial or total collapse if it has a masonry-block basement. In general, then, well-constructed wood-frame homes built on concrete slabs are the best bets. Masonry homes with concrete-block basements are the worst—again, if such factors as the distance from the epicenter, the soil type, and the nature of the surrounding buildings are equal.[15]

There are, however, steps that all prudent inhabitants of the Mississippi Valley can take in terms of safeguarding their homes *and* families in the event of even a moderately strong earthquake, though the steps will depend on what type of homes they live in:

• In a frame house, the frame should be adequately tied to the foundation, and the corners should be properly braced. (In many newer frame homes, the composition sheeting goes up to the corners of the house, providing less than adequate bracing during earthquakes. Ideally, the corners should either be braced with plywood sheets or diagonal bracing between the corner studs.) A contractor can probably best do this kind of work, though the work will not be inexpensive. There is, unfortunately, not much that can be done to reinforce the structure of a masonry home without spending a great deal of money.

• Basement walls should be reinforced with two-by-four studs and plywood or thick paneling. While this step will not stop the blocks from cracking, it can stop the blocks from cracking so badly that the walls collapse. Remodeling the basement is a dream many homeowners have, so this might be just the incentive to finally go ahead with such plans.

• A chimney should be no taller than it safely needs to be, and all the brick should be firmly in place. If the brick is loose and/or the chimney is unnecessarily tall, there is a much greater risk of falling bricks and subsequent injuries (or death) during an earthquake. The chimney column inside the home can also be reinforced. For instance, if the column is bare brick in the basement, it can be surrounded with studs and paneling at the same time the basement walls are being reinforced.

• A concrete block garage might also be reinforced to protect automobile(s) from falling blocks.

• Ideally, lightweight roofs are better than heavy roofs. Either way, the studs should be reinforced with sheets of plywood to prevent the whole structure from collapsing and possibly injuring those on floors below.

• The water heater, whether gas or electric, should be stabilized. Most water heaters are free-standing appliances, and they tend to "walk" during earthquakes and break free from their power source lines and waterlines, often causing unnecessary fires and/or floods. On the other hand, intact water heaters make excellent sources of pure water if a community's water mains are destroyed. (Shutting off the water coming into the home will assure that no contaminated water enters the plumbing system, and a family can live off the pure water in the water heater for days.) Metal straps or plumbing tape can be used to stabilize a water heater. The ends of the straps can be secured to nearby wall studs or other equally secure surfaces.

• Flexible gas hoses and connectors to gas-operated appliances (stoves, dryers, and so forth) should be installed. If such lines and connectors are rigid, they are liable to break during the shaking as the appliances are "walking" across the floor, and broken gas lines are one of the most serious problems during earthquakes. As an added precaution, the appliances themselves should be stabilized (anchored) so they cannot move across the floor.

• The electrical power shutoff switch, gas shutoff valve, and water shutoff valve should be made easily accessible. In addition, homeowners should know exactly where the switch and valves are located and keep any tools needed to turn them off (such as wrenches) nearby. If the shaking causes water, electrical, or gas line breaks, the homeowner will thus be able to quickly get to the source and stop any further flow into the home.

• Flammable and chemical substances should be stored low to the ground so they do not fall and break open and/or explode. They should be stored in strong containers and under strong surfaces so other objects do not fall on them and cause the same problems.

• Fire extinguishers and smoke alarms make good sense anyway, regardless of earthquake potential, so they should be installed and available. (Smoke alarms, however, may be of minimal value during an earthquake. If the tremor occurs at night and is strong enough to cause a fire, it will also undoubtedly be strong enough to wake the family. This is not always assured, though, so a smoke alarm may provide the first warning of the problem.)

• All large and/or heavy objects should be removed from high places. Even moderate jolts can knock such objects loose, causing them to come crashing down on someone's head. In fact, most injuries and deaths during earthquakes occur as a result of previously unsecured falling objects. This step is particularly important for families that have small children in the home. Not only will some children not know enough to take cover, but the distance from the object's original location to the child's head is greater than the distance to an adult's head, so the impact will be greater.

• Strong latches can be added to cabinets and other furniture that has doors. During earthquakes, unlatched cabinet and furniture doors often fly open, spewing their contents onto the floor. While the value of the objects is worth considering, the real reason for this step is, again, personal safety, especially the safety of children who might be underneath high cabinets or furniture doors at the time.

• Although it may dismay avid interior decorators, bolting or otherwise affixing tall furniture to the wall can also save lives and prevent injuries. Tall, heavy furniture is bound to come crashing floorward during hard jolts, and anyone underneath at the time will be injured, perhaps seriously. Bolts fixing furniture to the wall should be secured to the wall studs, not just to a plaster wall or composition sheeting. The studs will provide the bracing needed, unless the whole house collapses; bolts will rip out of plaster or composition sheeting. (Heavy pictures and mirrors can also be bolted to wall studs, for the same reasons.)

● Large, heavy, and/or valuable objects sitting on shelves should be protected by guardrails at the front of the shelves, so the objects do not fall. Of course, the shelves themselves should be securely braced and bolted to wall studs so they do not fall along with their contents.

● Heavy chandeliers, especially those with points, and heavy potted hanging plants can be killers during earthquakes. While it may not be necessary to replace such decorations, it is important to make sure they are secured well to the ceiling studs. While this may not make any difference in a powerful earthquake (where the whole ceiling may fall anyway), it can mean the difference between life and death during a moderate earthquake.

● Carpets can help to prevent furniture from "walking" across the floor during tremors. They can also absorb some (not all) of the shocks, thus possibly preventing tall, heavy pieces of furniture from falling. As with the basement remodeling, homeowners contemplating carpeting as a home improvement should know that it has an additional, more utilitarian, value.

● Outside, dead, dying, or otherwise weak trees should be cut down. They, too, can become killers during strong earthquakes if they topple over, strike people directly, cut power lines, and/or crash into homes.

● Finally, home tours are also recommended. As homeowners are making these tours, they should imagine fairy-tale giants lifting their homes and shaking them. Then they should ask themselves: What would be damaged by the shaking? In what ways could my family or I be injured or killed as a result of this kind of shaking? What can I do to prevent these damages, injuries, or deaths?

Homeowners should also prepare emergency supply kits to keep around their homes. They should imagine themselves and their families going camping for a few days—primitive camping with no utilities. Homeowners should put safely away in one place everything that would be required for such primitive camping. While homeowners may have everything needed to survive after an earthquake already in the house, it is important that the supplies be gathered together in one place so they can easily be accessed in case the earthquake severely damages the home, thus not allowing homeowners to go through their homes to find everything after the disaster. The recommended procedure: One should build (or buy) a large, solid box. It should be made out of one-inch-thick planks or one-inch plywood and be framed with two-by-fours. Such a box will stand a better chance of surviving if something heavy falls on it than will a flimsier container. The box should be stored somewhere the homeowner feels would be easy to get to after an earthquake, and should contain at least the following:

- flashlights (and extra batteries),
- a portable radio (and extra batteries),
- a first aid kit,
- a fire extinguisher,
- canned and dried foods,
- medication for people in the family who require it on a regular basis,
- a manual can opener,
- blankets and sleeping bags,
- heavy clothing for each family member,
- soap,
- candles and matches,
- a tent,
- an ax,
- charcoal, lighter fluid, and a grill (or camp stove),
- plastic utensils, plates, and cups.

The flashlights are important, because if there is a gas leak, lighting candles or turning on electrical switches (if the electricity is even working) can result in an explosion. Once the gas has been turned off and the area is safe, one can then use the candles.

Certain things should be kept separated from each other in the box. For instance, the lighter fluid should not be stored right next to the dried food or the matches. Perhaps small wooden compartments could be built inside the box to separate items and prevent them from falling or spilling.

Every few months, certain items should be taken from the box, used, and replaced with fresh supplies so they do not get stale. These include the batteries, the food (especially the dried foods), and the lighter fluid. The fire extinguisher should be checked occasionally, too, to make sure it is still effective. If it needs recharging, the local fire department can usually do the job for a small fee. Family members' medication, of course, should be changed more often, based on the expiration dates. Finally, the clothes, sleeping bags, and blankets should be aired out on occasion. In other words, while the "earthquake box" is designed for permanent storage of essential materials, it should be gone through regularly to make sure everything will be functional as it is needed.

Homeowners should also take steps to assure themselves of adequate postdisaster water supplies. Homeowners cannot rely solely on their water heaters to furnish such supplies. If a heater is leaking, or if contaminated water enters it before the valve is shut off, another source of fresh water will be needed. A few empty plastic milk jugs (or similar containers) should be filled with fresh water. These containers can be stored in the "earthquake box," but they will need replacing every month or so to keep them fresh. A better option is to store them in the freezer, where they will last longer without replacement. If they are stored in the freezer, they should still be replaced about once a year or so.

One of the hallmarks of modern society is the fears our children have. A survey of children's fears conducted in 1935, for example, listed animals, dark rooms, strangers, loud noises, and loneliness as the top five, whereas a survey conducted in 1982 listed death of a parent, parents with cancer, divorce, nuclear war, and poor grades as the top five.[16] Obviously, the media have heavily influenced youngsters' fears today, but in addition, parents now seem more prone to pass their own worries on to their children than they were fifty years ago, when children were allowed to be children and did not have to take on adults' problems. Parents who believe that children have enough to worry about on their own without having to hear about and take on the parents' problems will want to discuss the earthquake problem very carefully and sensitively with their children. In other words, the purpose of discussing the topic with children is not to alarm them or otherwise frighten them; it is only to let them know how important it is to take action when it occurs and what that action should be.

Parents and children should discuss the safest place to be in each room of their home during an earthquake. It is important to have a place in each room, because time will not allow anyone to run from room to room to find a place of safety. An earthquake occurs so quickly, occupants will be lucky to get under something right where they are. Appendix 3 contains the necessary instructions on what to do during and immediately after an earthquake. This information can be used for family earthquake drills.

Readers who feel overwhelmed by the above information may simply want to consult this encapsulated list of the most important things to do, which even collectively require a minimum amount of time and investment:

- Purchase earthquake insurance.
- Secure the water heater.
- Know where the electric, gas, and water turn-off switches/ valves are.
- Bolt tall furniture and remove heavy objects from high places.
- Install flexible gas hoses and connectors.
- Build and fill an "earthquake box" with necessities.
- Know what to do when an earthquake occurs.

7. Surviving the Next Major New Madrid Earthquake

The Coming Cataclysm

A loud booming noise is the first sign something is amiss. As the earth resounds like a giant loudspeaker, flashes of light emanate from the ground, and the ground begins to shudder. Within seconds, the shaking becomes violent, and giant fissures rip open, oblivious to what is in their paths. Geysers of wet sand, water, and rotting debris vomit forth from deep within the earth. The land begins to sink, slowly working its way into the wet sand below. There is a heavy sulphurous stench, and the air blackens with soot, dust, and dirt. Breathing becomes difficult.

Bluffs give way, and thousands of tons of rock fall with ear-splitting reverberations. Giant masses of land cease clinging to hillsides and roll downward, burying whatever is in their way. Within a radius of one hundred miles from the epicenter, the soil in low-lying areas and along rivers shakes like jelly, losing its cohesion. Forests for hundreds of miles crackle, and the trees snap and fall like hundreds of thousands of blades of grass being mowed and sheared at ground level. The giant waterways of the region, the Mississippi and Ohio rivers, boil with vengeance, rise, and overflow their banks with raging floods. Dams crack and eventually break, loosing torrents of water onto the structures in their way.

Houses and buildings in the region shake, shudder, and begin to collapse as the ground beneath them jars and jerks their frames mercilessly. Those designed to sustain earthquake vibration strain to hold their forms, and some succeed. Others break apart in sections. Still others are demolished as nearby structures not built to withstand the shaking come crashing down on them. The sounds of panes of shattering glass echo between the falling buildings, as the rubble piles up, and dust billows around the scenes of devastation. Wood snaps, and frame structures become unrecognizable piles of kindling, laced with wire, plaster, pipes, and glass. Masonry school buildings fall one by one, as do gymnasiums, churches, courthouses, theaters, and malls. High-rise buildings drop like accordians, layer upon layer smashing on top of one another, until the sheer weight collapses the structures completely.

Emergency service units attempt to mobilize. The effort is futile: there is nothing left of their headquarters. Hospitals collapse, burying those inside—patients and staff alike. Communities near the epicenter and along nearby river channels are in total chaos. Those farther away and on higher ground suffer major damage but remain partially functional. The rest of the nation is aware of the event almost instantaneously as the ground from the Atlantic Ocean to the Rocky Mountains shakes frighteningly. People in major metropolitan regions hundreds of miles away feel the ground shake and watch the buildings sway precariously. Highways buckle, crack, and wave uncontrollably. Some automobiles, trucks, and other vehicles careen and crash into one another, while others drop into fissures ripping across the landscape. Bridges collapse almost instantaneously.

Rail lines twist like spaghetti and rip away from the thick ties embedded in the ground, which snap like toothpicks. The undulating ground, liquefied soil, and fissures leave little of what were once straight bands of gleaming steel. Rail cars and their engines fall over like model toys, spilling their contents: the region is littered with thousands of gallons of hazardous chemicals. Railroad bridges collapse like their highway counterparts, crashing into rivers and deep ravines below. Barges on the Mississippi, Ohio, and Missouri rivers lose control. Many are catapulted ashore, where they crash into the riverbanks. Other boats capsize, taking with them their occupants.

The liquefaction, fissures, and thundering ground waves break the network of underground gas lines that snake through the Mississippi Valley on their way across country. Major fires erupt as the gas spews into the atmosphere. Cities in the northeastern United States are cut off from their sources of power. Gas lines serving regional communities break as well. Power plants in the region shudder and begin to crumble, and generating equipment short-circuits. The major power lines criss-crossing the Midwest break and snake earthward, showering the ground with sparks and setting innumerable fires. The giant metal towers supporting them lose stability and bend as though made of wire. Water mains break throughout the region, and millions of gallons of fresh water seep into the soil, as fires spread uncontrollably because there is no water to fight them with. Sewer lines break, releasing all of their contents into the ground. The stunned citizenry and emergency services personnel attempt to communicate with one another over communication lines, but to little avail. Telephone lines are down. Radio and television transmitters are no longer standing. The region is effectively without power, and everything that depends on it is unceremoniously and unexpectedly shut down. Communities are left to fend for themselves—isolated from one another and from the outside world.

Factories and other industrial complexes are destroyed. Banking and all other forms of commerce come to an immediate halt. Dozens of years worth of valuable records stored on computer disks and tape are obliterated, never to be recaptured. Underground tanks filled with thousands of gallons of fuel rupture and explode at gas stations. Hazardous waste dump sites in the region are battered, their contents leaking permanently into the ground. The underground coal mines collapse, burying all within. Farming, the region's major form of economic sustenance, is destroyed, as the rich farmland takes on the appearance of the terrain after a nuclear holocaust.

The Mississippi Valley civilization, two hundred years in the making, is obliterated—in a matter of mere minutes.

Magnitudes, Intensities, and Damage and Casualty Figures

It is impossible to say just how powerful the next major New Madrid earthquake will be. In fact, the formal studies conducted on damage and casualty potentials from a major earthquake in the Mississippi Valley perform their calculations using a number of different possibilities. A technical work published in 1981 estimates damage and casualty figures for earthquakes ranging in Mercalli intensity from VIII to XII[1] (with estimated magnitudes of approximately 6.5 to 8.7). Another study reports calculations based on an 8.6 magnitude earthquake (maximum intensity XI), and yet another makes two estimates—one based on a 7.6 magnitude earthquake, the second on an 8.6 magnitude.[2] Currently, according to Otto Nuttli, there may be enough energy stored in the New Madrid Seismic Zone to produce a magnitude 7.6 earthquake. If all of this energy were released at once, in other words, a 7.6 earthquake would rattle through the region. If only part of the energy were released, the magnitude would be less.

The scenario described at the beginning of this chapter is for an estimated magnitude 8.5 earthquake. The results of milder shocks—a 6.5 or a 7.5—would of course be different. It is important to remember in this regard the exponential nature of the Richter magnitude scale (see table 6 in chapter 3). While a 6.5 magnitude earthquake would release energy equivalent to the explosion of approximately thirty thousand tons of TNT, a 7.5 would release energy equivalent to approximately nine hundred thousand tons of TNT, and an 8.5 would release energy equivalent to approximately 30 million tons of TNT. In other words, an 8.5 magnitude earthquake is almost nine hundred times more powerful than a 6.5 magnitude earthquake. To put the relationship in even sharper perspective: the November 9, 1968, Mississippi Valley earthquake was a magnitude 5.5 event

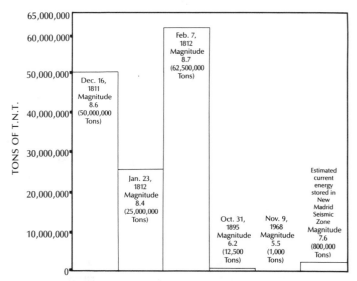

Chart 3. Relative Power of Historical Mississippi Valley Earthquakes

(equal to the explosion of approximately one thousand tons of TNT); the February 7, 1812, New Madrid "hard shock" was a magnitude 8.7 event (equal to the explosion of approximately 62.5 million tons of TNT). That is, as can be seen on chart 3, the second 1812 earthquake was over sixty-two thousand times more powerful than the 1968 earthquake.

Other variables that must be considered are the location and depth of the earthquake. If the quake occurs in the southern end of the New Madrid fault zone, near Memphis, damage and casualties will be much greater than if it occurs in a more rural region of the fault zone. If it is a shallow earthquake (one or two miles deep), there will be a great intensity near the epicenter, but the intensity will fall off rapidly at distances. On the other hand, if it is a deep earthquake (three to twelve miles deep), the area it damages will be greater, because the energy will spread out over greater distances. (See chapter 3 for an explanation of the dynamics involved.)

The time of the event is important, too. If the earthquake occurs at night, when most people are asleep in their frame homes and traffic is light, casualties will be substantially less than if it occurs during a weekday, when schools and masonry buildings are full and the highways are busy.

Taking all these factors into account, it becomes obvious that it is currently impossible to actually predict the casualty figures for a major earthquake. The 1971 Peru-Ecuador earthquake (magnitude 7.5), for example, killed one person and injured three. The 1976 Guatemala earthquake (also magnitude 7.5) killed twenty-three thousand people and injured over seventy-six thousand. The 1972 Taiwan earthquake (magnitude 6.9) killed four people and injured eleven. The 1980 southern Italy earthquake (magnitude 6.8) killed almost three thousand people and injured nearly nine thousand.[3]

The three accompanying maps (maps 3, 4, and 5) show estimated intensities for three earthquakes measuring, respectively, magnitude 6.5, magnitude 7.5, and magnitude 8.5.[4] To determine approximately how much damage a community or area is estimated to suffer, the reader can study the map for each magnitude, find the community's location on the map, read the accompanying Mercalli intensity number (the roman numeral), and then read the description for that number in appendix 1. For example, the estimated intensity figure in Paducah, Kentucky, for a 6.5 magnitude earthquake is VII. The interested reader would therefore read the description for a VII intensity earthquake in appendix 1.

A 6.5 Magnitude Earthquake. An earthquake of 6.5 magnitude could have severe consequences if it happened directly under one of the cities in the zone, suggests Arch Johnston, but "if it happened in a rural area with low population and fewer buildings, it would be less severe." Otto Nuttli agrees that an earthquake of 6.5 magnitude would probably cause few deaths unless it occurred near a city: "With the unreinforced masonry buildings and poor soil conditions in Memphis, for instance, an earthquake there could be catastrophic."[5]

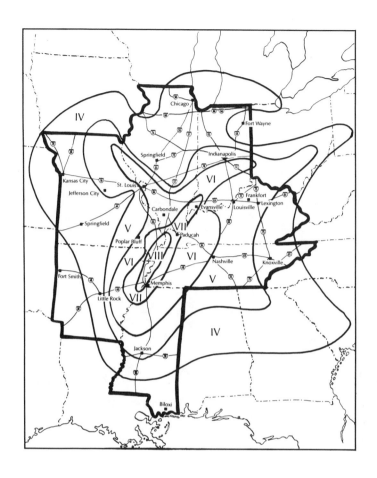

Map 3. Estimated Intensities from a 6.5 Magnitude Earthquake

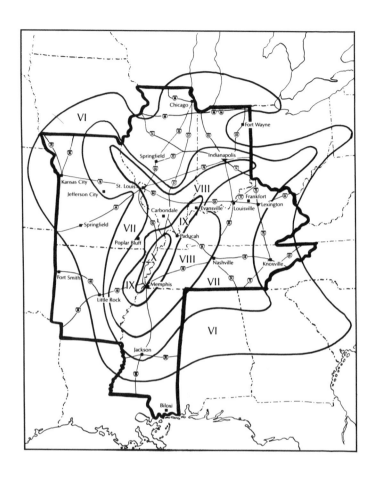

Map 4. Estimated Intensities from a 7.5 Magnitude Earthquake

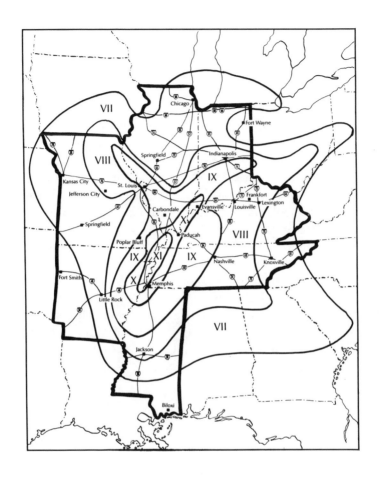

Map 5. Estimated Intensities from an 8.5 Magnitude Earthquake

A 7.5 Magnitude Earthquake. With an earthquake of 7.5 magnitude, "we could have [1985] Mexico City-type damage as far away as St. Louis," warns Nuttli. At greater distances from the epicenter, the amplitude of the waves would not be as great, but the shaking would last longer, which could cause more damage. These low-frequency, long-shaking waves could cause problems to high-rise buildings in metropolitan areas. Such buildings would not collapse, but they would sustain a great deal of interior damage. A 7.5 might be survivable, though, in terms of the economic and social viability of the area. Lawrence Malinconico feels that "the area would be able to pick up and take off afterwards."[6]

According to the Allen and Hoshall (1985) FEMA report, a 7.6 magnitude earthquake would kill approximately 300 people in six major cities in the Midwest (Memphis, Paducah, Carbondale, Evansville, Poplar Bluff, and Little Rock) if it occurred at night; 3,000 if it occurred during the day, approximately 750 of these being school children. The estimated restoration and replacement costs for structures and utilities in the six cities surveyed would be approximately $38 billion.[7]

An 8.5 Magnitude Earthquake. Allen and Hoshall predict almost 700 killed in the six cities if an 8.6 magnitude event occurred at night; almost 5,000 if it occurred during the day, approximately 1,350 of these being school children. The estimated restoration and replacement costs for structures and utilities in the six cities surveyed would be approximately $52 billion.[8]

"There would be no other natural phenomenon that would cause such widespread damage and affect so many people's lives," states Nuttli. "It would tax the resources of even our federal government, and would certainly be beyond the capabilities of our state and local governments." An 8.5 would seriously affect the economic and social viability of this area for a long time afterward, agrees Malinconico. "I don't think people realize what would happen. Actually, the region could probably not recover from it."

Socioeconomic Results

Medical Needs

If the Midwest experiences a major earthquake, medically it will be under battlefield conditions, called "triage," according to Dr. Richard Moy.[9] "Even if the hospitals survive," Dr. Moy contends, "they will be presented with so many different people and such an array of problems, that they won't be able to cope alone." The ability to practice civilian medicine, including emergency trauma medicine, will be overwhelmed.

Under these battlefield conditions, the first step medical personnel will need to take will be to look for those who are seriously injured and need care immediately, before they lose life or limb. These people will need medical assistance to stop their bleeding or otherwise stabilize them. Because more people will be coming in, there will be no opportunities to conduct a time-consuming operation on anyone. "If you do, you may have great success with that one person, but in the meantime three or four other people will have died waiting," Dr. Moy warns. A second group of people, those who are "shell-shocked" but not seriously wounded, will have to be sent elsewhere—possibly, for example, given tranquilizers and sent to a local shelter—to allow room for those who need immediate emergency treatment. It may be necessary to stabilize a third group (those who are so badly injured they may not survive—people with extreme burns over 60 or 70 percent of their bodies, for example), give them blood or morphine, and then place them in a shelter, hoping that six or seven hours later when the first group has been cared for the people in this third group will still be alive. Since this third group will require such intensive care, it will not be practical to tie up the whole system for them. "The first group *must* be the priority," Moy emphasizes.

Individual Responses, Group Results

Following disasters, most people appear stunned for a short period of time. They, however, react calmly and rationally. "By and large, people everywhere respond well to disasters; panic rarely occurs," states Dennis Mileti. Panic results only when four conditions exist. For panic to result from a fire in a theater, for example: (1) The fire must convince most people that they will die unless they get out. "And it's very hard to convince someone he or she is going to die," adds Mileti. (2) There must be escape routes that, if traversed, can save the people's lives. If there are no escape routes, there is not likely to be panic. (3) The escape routes must be temporarily closed off. (4) It must be obvious to the people that if they do not move quickly, they will not be part of the 20 percent or so who will be able to save themselves, but instead will be part of the 80 percent who will die because of not moving quickly enough. "If these four conditions exist, then even the most frail, loving person will step on your face to get out," adds Mileti. Such instances, while they have occurred after earthquakes, are very rare.[10]

Another heartening fact is that, while tradition holds that disasters cause social cohesiveness to come apart, they actually strengthen the bond between people. Most people do not become selfish and steal from one another. The opposite occurs. Civilization actually strengthens; people become more personal with each other and more compassionate. While few people talk to strangers under ordinary circumstances, they do so following disasters. "People actually revert back to the more ancient forms of communication. They revert from the 'big city' habit of ignoring people to a more rural type of friendliness," suggests Mileti.

Small, informal groups of strangers band together after disasters and go around rescuing those who need help. "In fact, in most disasters, victims are rescued by other victims," continues Mileti, "not by outside help or heavy equipment." By the time the media arrive to shed publicity on disasters, these small, informal groups have broken up and outside help has taken over. "The majority of people are rescued by friends, family, and neighbors," agrees E. L. Quarantelli. "Outside emergency groups arrive too late to help the majority of people."[11]

Surprisingly, looting is almost completely nonexistent following disasters, according to Mileti. Looting never occurs after earthquakes, he insists. There are stories about looting after every disaster, Mileti feels, because people expect it to occur. Police put up barricades to stop looters and claim how successful they are, but they are so successful, he points out, because there are no looters to stop in the first place. On occasion, the police or media may notice someone digging through the rubble of a home or building and assume it is a looter. After the person is questioned, however, they discover the person owned the home or building and was looking for possessions. Sociologists, in fact, have tried to interview arrested looters after over one hundred disasters, according to Mileti, but they can never find any. There are simply none of them in jail. It seems, then, that while looting may be standard procedure during and following riots and other civil disturbances, it is nothing to be concerned about following disasters. "I have developed an undying view that the public is a resource, not a problem, following a disaster," emphasizes Mileti. "Government officials too often think of themselves as parents who must help the helpless public in emergencies. Immediately following a disaster, this simply just isn't the case."

The "Second Disaster"

The one or two days following disasters, as described above, are marked by periods of cooperation and selflessness. Within a week, however, these quickly deteriorate to conflict and selfishness. Outside emergency organizations often launch power struggles among themselves and with local emergency organizations, each jockeying for position, power, and status. That many of the people working on the relief effort are insufficiently trained and simply get in the way only exacerbates the problem.

Following disasters, public shelters are grossly underused since people prefer to stay with relatives. After a few days, however, the relatives tire of them, conflicts arise, and the victims go looking for aid from governmental bodies and other emergency groups. Since these organizations are in such disarray, though, victims fail to get what they feel they deserve, and tempers rise. People become impatient with bureaucratic procedures, such as applying for temporary housing. "After the esprit de corps has evaporated, people start shouting at public officials," states Frederick Ahearn.[12]

The media, which may have been quite helpful prior to the disaster, can also become a major problem. "Their role is to come in afterward and look for the gruesome stories," notes Ahearn. Their norms are no longer the same as the community's. Following a disaster, furthermore, the media are in competition with each other to find the dramatic and play it up. "Some of what they do is totally unnecessary—going up to people whose homes are destroyed and asking them how they feel," adds Quarantelli. "You can bet they will have a victim break down and cry in front of the camera, and this is the one they will show. It bothers a lot of people, and quite correctly, that the media have created this unnecessary stress on disaster victims."

The media can be a particular problem in hospitals, where critical procedures are taking place. "Newspaper and radio reports aren't really the problem; it's the TV people," complains Michael Ertle, manager of safety and security for Little Company of Mary Hospital (Torrance, California), who specializes in disaster security procedures. "TV people want blood! They will barge in anywhere they want with an entourage of cameras and microphones, disrupt work in the emergency room, and demand admittance to patients' rooms. They need visuals for their newscasts, so they want to get whatever they can. If someone tries to stop them, they'll put lights and microphones on them, roll the cameras, and demand 'What is your name?' They try to be very intimidating." Ertle emphasizes that hospitals are private property and that the staff has every right to refuse entrance to people—including the media—and to control their behavior if they are allowed inside. Of course, unless and until the public curbs its seemingly insatiable appetite for viewing human suffering, these problems will continue to occur. In fairness, however, Ertle adds that the media can play a very important role in disaster coverage. If there is enough exposure around the nation to the problem, it helps greatly with getting federal assistance. "The more publicity a disaster generates, the better the chance of outside aid," he adds.[13]

Changes Over Time

In the months following a major disaster, families (both nuclear and extended) tend to grow closer, having gone through the problem together and having relied more on each other than they had ever expected needing to. After an earthquake of less than 8.0 magnitude or so, communities, unlike families, have one of two ways of going. They can succeed in rising above the disaster and become more tightly knit, or they can fail to overcome their problems and suffer long-term consequences. (To increase chances of success, community officials should help provide accurate information to citizens, instill a sense of hopefulness in the community, remind people that they have gone through the disaster together and that they should be proud of themselves, and exhort the community to move forward together in a new spirit of progress.) In either case, three years after a disaster, its effects will still be felt. Prices will still be dramatically inflated, especially those of building materials, and political controversy among various groups about what and where to rebuild will still occur. Ten years later, however, there may be no sociological or economic negative effects. People may have succeeded in rebuilding (due to the influx of outside aid), and all the old economic and social systems may be back in place. In fact, many communities become more socially and economically healthy after disasters because of the new influx of money.

If, on the other hand, an area were to experience an 8.5 or stronger cataclysm, total despair may well reign. "The information we have is for earthquakes that affect just one or two communities, and where the rest of the nation is in a position to help," admits Mileti. "We're not talking about the kind of earthquake that would exceed the ability of the whole nation to help." And even moderately strong earthquakes can eventually result in long-term despair if they are followed by strong aftershocks. After a major earthquake in Italy in 1976, for instance, the citizens rose to the occasion and began rebuilding with a spirit of optimism. Following a second powerful tremor four months later, however, most of them lost their spirit, and many left the area.[14]

Following the three major New Madrid earthquakes in 1811–1812, all of which exceeded 8.0 magnitude, most inhabitants immediately deserted the region, and over the next ten years, as the aftershocks continued, even more left. It would be interesting to study this phenomenon in detail to see what impact this early historical exodus had on the current socioeconomic conditions in the region. It is possible that the Mississippi Valley would be a major economic region with many more large cities than it now has had so many early settlers not left the region almost two hundred years ago, and had so many other settlers not stayed away from the region because of its reputation. What this says about the future shape of the Mississippi Valley if it ever has another disaster approaching in immensity those of 1811–1812 can only be guessed at.

Psychological Reactions to Major Disaster

On February 26, 1972, a dam burst and flooded Buffalo Creek Valley, West Virginia, leaving 125 dead and 4,000 homeless. Five hundred homes were destroyed, and fifty million dollars in property damage was sustained. Within days, the psychological problems began. Insomnia, fear of loud noises, survival guilt feelings, amnesia, eating disorders, family hostility and resentment, apathy, pervasive feelings of hopelessness, isolation and withdrawal, fear, death anxiety, nightmares, depression, belligerence, alcohol abuse, inability to concentrate, blank spells, and indecisiveness persisted with many of the residents for weeks, months, even years. The Buffalo Creek disaster, a nightmare for its residents, was a dream for disaster sociologists. Rarely had they ever had the opportunity to study the reactions of people who survived such a total, intense, and thoroughly devastating disaster. The Buffalo Creek flood has become one of the most studied disasters in history. Unlike most disasters that destroy a portion of a community, the Buffalo Creek flood left its impact on everyone there. A whole community was destroyed, and the resulting sociological and psychological impact on the survivors has told social scientists a great deal about how people react to major devastation.[15]

General Psychological Reactions

Immediately following disasters, people tend to respond quite well. The events seem to have a narcotic effect on survivors, temporarily preventing them from grasping how much their lives and environments have changed. To cope with the changes, people tend to associate the unfamiliar with something familiar and thus lend some continuity to their otherwise disrupted lives. For the first few hours following a disaster, victims tend to remain quiet, undemanding, pensive, and helpful.[16] Many go into a state of shock, express apathy, and seem unable to act or make decisions. They can discuss the most horrifying and gruesome scenes and experiences with seeming dispassion. Many also engage in useless activity, trying to keep busy with meaningless tasks, such as digging through rubble without purpose.[17]

In terms of numbers, approximately 75 percent of such victims seem stunned and bewildered; approximately 10 percent to 15 percent seem confused, paralyzed by anxiety, or hysterical; and the other 10 percent to 15 percent remain collected and in control.[18] To help see them through the catastrophe, many people rely on humor, morbid and otherwise. Most direct it at themselves and their experiences, not at others, because the spirit of cooperation is still strong among the survivors.

Within a few days, however, psychological problems can develop. The more horrifying, devastating, and widespread the disaster, the more serious and pervasive the reactions. Some common psychological problems, as mentioned above, are exhaustion, depression, confusion, paralyzing anxiety, a tendency to cry, apathy, irritability, anger, withdrawal, insomnia, guilt (at having survived while others did not), increased alcohol or drug use, and family conflict. Physical problems include backaches, headaches, loss of appetite, and exhaustion. Not all persons experience all of these symptoms, though. Most people experience one or more symptoms mildly and work them out by constantly talking about the disaster—a healthy psychological defense mechanism. In fact, talking out problems with another victim is often as useful as talking with a psychiatric social worker.

Some people, however, especially those who are particularly hard hit by a disaster (those suffering the loss of home or loved one, for example), experience some serious problems that require professional assistance. While most people want to talk about their experiences, about 20 percent *need* to talk about them. These people are victims of what is often called "disaster syndrome."[19] In most cases, though, unless the problems are really serious, such people eventually work through their emotional problems and resume healthy lives.

Within a few months following a disaster, other problems can manifest themselves. It is not uncommon to find increases in socially unacceptable behavior (robbery, juvenile delinquency, and drug and alcohol abuse.) Family problems can also increase, with spouses feeling that their mates do not understand them. (With both spouses needing to talk more than listen, conflicts and resentment naturally build.) Other relationships can change (among extended family members, friends, coworkers, and employers, for instance) as victims attempt to readjust their lives and assume new responsibilities (such as having to find new work or raise children alone). In addition, it is not uncommon to find victims who have had to relocate still grieving over the loss of their homes for a year or more after losing them.

In terms of solutions, it should be emphasized that it is alright for people to feel what they are feeling. "It's OK to feel lousy and depressed; almost everyone does," explains Frederick Ahearn. "You've experienced a loss and an upheaval." Ahearn recommends that victims talk out experiences and concerns with anyone who will willingly lend a listening ear—friends, family, coworkers, neighbors, ministers, or professional counselors. He adds that the media can perform a major service to their readers, viewers, and listeners by doing stories on victims a few months after a disaster. A story on what it is like for the average citizen to go through a disaster can help others identify with the feelings and realize that they are coping adequately. After the 1972 Nicaragua earthquake, for instance, the local newspaper did a series on how the survivors were the real heroes of the disaster. "It was excellent, because it raised everyone's morale," adds Ahearn. The newspapers helped victims understand that it was alright to cry, to feel depressed, and to be disoriented. It helped them feel good about themselves and their community.

Special Groups, Special Problems

There are a number of groups who may, because of their situations, experience more trauma from the disaster than others. One such group consists of people who were experiencing major life problems just prior to the disaster: those who are recently widowed or divorced, those who have just changed jobs, those who have recently had major surgery or who have just moved, for example. "For these people, the disaster may just be the 'straw that breaks the camel's back,'" according to Ahearn.

Another group consists of people who have moved to the community shortly before the disaster. Not only do they experience the stress of moving and then the stress of the disaster, but they experience the pain of isolation after the disaster, since, unlike the other citizens in the community, they often do not have long-time friends, neighbors, and extended family members immediately at hand to talk with and rely on.

A third group that may experience problems related to the disaster are those who were not even *in* the disaster. These are the families of disaster victims and survivors. The greatest hardship, of course, is on those who have lost family members in the disaster. There is also a great deal of hardship, though, on families who must wait hours, even days, to get word from the disaster scene on whether or not their spouses, children, or parents survived.

A fourth group consists of those with a history of chronic mental or emotional problems prior to the disaster. During the 1971 San Fernando earthquake, there was no hysteria among patients at a mental hospital. In fact, for a short period of time, those diagnosed as psychotic seemed to become temporarily rational. Within a few hours, however, they reverted to their original pathologic states.[20] Over the long term (weeks and months following a disaster), readmission rates to mental hospitals tend to rise, though new admissions do not show any marked increase.[21] Thus, it seems that those with psychiatric problems before a disaster need more help following a disaster, but those without such problems before a disaster rarely develop them because of the disaster.

There are three *major* groups in the community whose reactions to disaster must be looked at closely: (1) the children, (2) the elderly, and (3) the "helpers" (such as volunteer disaster workers, medical personnel, and psychiatric counselors).

The Children. There is little question that children experience psychological problems following a disaster to a much greater and more serious degree than do adults. Common problems include sleeping and eating disorders, fear of sleeping alone, behavioral problems, fear of leaving parents' sides, phobic reactions, withdrawal, belligerence, irritability, resumption of earlier speech habits ("baby talk"), increased sibling rivalry, thumbsucking, loss of interest in toys and hobbies, and generalized fear.[22]

There are a number of causes for and solutions to these problems. Often, parents are so busy getting their own lives reorganized that they either forget, or do not feel they have time, to attend to their children's special needs related to the disaster—both physical and psychological needs. The children, in many cases, feel they are essentially left to fend for themselves, even though their parents may be taking care of their basic needs (food and shelter, for example). These children need to talk about the disaster, to face and express their fears, but often do not feel they have the opportunity. Other children become extremely bothered by their parents' reactions to disaster. It frightens them when they see their parents going through their own emotional problems related to the disaster. In both cases, the love, strength, information, understanding, and encouragement that they had come to rely on so heavily from their parents in the past are no longer there. As Ahearn emphasizes, "Children often aren't capable of understanding what is going on, and they don't have an adult's ability to cope with the stress."

It is vitally important that parents either help their children work through their concerns or see that they get professional help if they are unable to meet these needs themselves. Parents who tackle the job themselves should encourage their children to discuss their problems, listen attentively and supportively, and allow regressive (clinging) behavior for a period of time until the children seem to be able to function again on their own. Group settings also seem to be productive in getting children to discuss their fears, concerns, and anxieties. When children hear other children relating their problems, they feel more comfortable relating their own and understand that others have been going through exactly the same problems. Following such sessions, many parents have been surprised that their children had actually harbored such strong anxieties.

The Elderly.　There seems to be some disagreement as to how the elderly respond to disaster. Some experiences suggest that they respond fairly poorly. Many of the elderly live alone, and since most of their families and friends have passed away or left, they feel isolated to begin with. All they have left are their homes and possessions. If these are taken from them during a disaster and they (the elderly) are forced to relocate, it is not unreasonable to assume that they will experience a second disaster, the one of relocation following the first disaster. Many may develop feelings of despair, depression, and hopelessness. In addition, many of them have been self-reliant all their adult lives, and either do not know where to go or do not want to go for assistance of any kind. On the other hand, in some instances the elderly have seemed to rebound quite well from disasters. They report few problems of an emotional nature and simply seek more pragmatic assistance (such as housing, loans, and food). "It's difficult to predict," admits Ahearn. "In many cases, the elderly have more at risk, because they may have fewer resources to deal with the aftermath of a disaster. And given their age, they may have less energy to want to 'pick up the pieces' and start over again. On the other hand, because of their life experiences, they may actually be more capable of dealing with disaster than some younger people."

The "Helpers." There are many different groups of people and organizations who assist after a disaster. All of them are potentially at risk for ensuing psychological problems if they do not take proper precautions and do not seek special assistance themselves after the disaster itself is over.

One subgroup of helpers is made up of the volunteers who actually help in rescue operations. "If these people are exposed to a lot of gore and other horrifying scenes, it's not unlikely they could benefit from some kind of counseling afterwards," states E. L. Quarantelli. These firefighters, police officers, National Guard personnel, and others often work around the clock with little sleep. They also see a lot of horror—often having to, for example, pick up parts of bodies. Besides having to deal with the problems such horror causes, many members of this group often feel guilt at not having arrived earlier to help or at not having done even more than they did to help.

A second subgroup of helpers is made up of the medical personnel who also often work around the clock to assist the injured. They frequently suffer experiences and quandaries similar to those of the actual rescuers: Should I make sure my own family is alright, or should I stay here and help others? How much attention should I give this person, when there are twenty more waiting for assistance? Should I work continuously, or take a break on occasion? How many will die if I do not work continuously? Should I continue working even though my own safety may be in jeopardy from aftershocks, falling debris, and dangerous weather conditions?

A third subgroup of helpers consists of psychiatric counselors who meet with victims during the days, weeks, and months following a disaster. Often, they have been victims themselves and feel a strong need to get their own lives in order before taking on anyone else's problems. Further, as they hear the same stories over and over again, they can lose perspective, begin to feel helpless and frustrated, and wonder about their effectiveness. Burn out, frustration, fatigue, irritability, impatience, and ambivalence, and a feeling of being overwhelmed are not uncommon among this subgroup.

Experts suggest that all disaster workers should get proper counseling themselves (1) to keep themselves functional and (2) to allow them to remain effective for those who need them.

Post-Traumatic Stress Disorder

On occasion, disaster victims experience delayed reactions to the horror they were exposed to. Some begin to exhibit signs that something is wrong six months after the disaster. Others do not show the effects until a year, five years, or ten years after. In many, the problems last a lifetime if not treated. Tell-tale signs include recurrent recollections or dreams of the event, sudden action as if the event were reoccurring, numbing of responsiveness to the external world, difficulty in interpersonal relationships, diminished interest in one or more significant activities, lack of emotion, feelings of detachment and estrangement from others, sleep disturbance, vacant expressions, survival guilt, and an inability to concentrate. Victims often also experience marital conflicts serious enough to lead to divorce and the inability to hold a job.

For decades, this problem was called "shell shock" and was thought to be limited to soldiers returning from war. The condition was first noticed after World War I, but by the 1940s it was obvious that victims of any major traumatic event (such as a military action, hostage situation, auto accident, heart attack, violent crime, plane crash, or natural disaster) could be victims of what was by then called psychic trauma. Today, this condition is clinically recognized and identified as Post-Traumatic Stress Disorder (PTSD). Unfortunately, the term has been bandied about loosely since being introduced in 1980. Actually, very few victims of traumatic events ever genuinely experience PTSD. "A small subset of people may actually experience PTSD," cautions Mileti. "I would be surprised if even as many as 1 percent of the population, for instance, would be affected by PTSD after an earthquake."

There will probably be, nonetheless, at least some survivors who suffer from PTSD. "If the disaster has a high horror factor, such as a lot of bodies, blood, and other gruesome scenes, exposed victims may develop PTSD," states Ahearn. "Examples would be seeing a loved one crushed to death, losing your own leg, and so forth," adds Mileti. In addition, if the experience is prolonged, such as with an earthquake where victims are trapped for days, the potential becomes greater for PTSD. Unfortunately, children are just as susceptible to PTSD as are adults.

Overcoming PTSD requires:

1. identification of the problem,
2. admission of the problem by the victim, and
3. counseling.

There can be, in other words, actually three disasters related to a major earthquake: (1) the earthquake itself and the ensuing physical damage; (2) the subsequent disruption of the community (including relocations, job changes, and rebuilding costs); and (3) the later psychological trauma that, if not addressed promptly and efficiently, can leave permanent scars on the survivors.

The event may, though, in its own way, leave some positive reminders. Trauma can help a person grow—to become stronger, to become a better person, to be better able to cope with future problems he or she may encounter.

APPENDIXES
NOTES
BIBLIOGRAPHY
INDEX

Appendix 1
Modified Mercalli Intensity Scale of 1931

The following is a reprint of "Modified Mercalli Intensity Scale of 1931 (Adapted from Sieberg's Mercalli-Cancani Scale, Modified and Condensed)," *Earthquake Information Bulletin*, July-August 1977, 30–31.

I

Not felt—or, except rarely under especially favorable circumstances. Under certain conditions, at and outside the boundary of the area in which a great shock is felt: sometimes birds, animals, reported uneasy or disturbed; sometimes dizziness or nausea experienced; sometimes trees, structures, liquids, bodies of water, may sway; doors may swing, very slowly.

II

Felt indoors by few, especially on upper floors, or by sensitive or nervous persons. Also, as in grade I, but often more noticeably: sometimes hanging objects may swing, especially when delicately suspended; sometimes trees, structures, liquids, bodies of water, may sway, doors may swing, very slowly; sometimes birds, animals, reported uneasy or disturbed; sometimes dizziness or nausea experienced.

III

Felt indoors by several, motion usually rapid vibration. Sometimes not recognized to be an earthquake at first. Duration estimated in some cases. Vibration like that due to passing of light or lightly loaded trucks, or heavy trucks some distance away. Hanging objects may swing slightly. Movements may be appreciable on upper levels of tall structures. Rocked standing motor cars slightly.

IV

Felt indoors by many, outdoors by few. Awakened few, especially light sleepers. Frightened no one, unless apprehensive from previous experience. Vibration like that due to passing of heavy or heavily loaded trucks. Sensation like heavy body striking building, or falling of heavy objects inside. Rattling of dishes, windows, doors; glassware and crockery clink and clash. Creaking of walls, frame, especially in the upper range of this grade. Hanging objects swung, in numerous instances. Slightly disturbed liquids in open vessels. Rocked standing motor cars noticeably.

V

Felt indoors by practically all, outdoors by many or most; outdoors direction estimated. Awakened many or most. Frightened few, slight excitement, a few ran outdoors. Buildings trembled throughout. Broke dishes, glassware, to some extent. Cracked windows, in some cases, but not generally. Overturned vases, small or unstable objects, in many instances, with occasional fall. Hanging objects, doors, swung generally or considerably. Knocked pictures against walls or swung them out of place. Opened or closed doors, shutters, abruptly. Pendulum clocks stopped, started, or ran fast or slow. Moved small objects, furnishings, the latter to slight extent. Spilled liquids in small amounts from well-filled open containers. Trees, bushes, shaken slightly.

VI

Felt by all, indoors and outdoors. Frightened many, excitement general, some alarm, many ran outdoors. Awakened all. Persons made to move unsteadily. Trees, bushes, shaken slightly to moderately. Liquid set in strong motion. Small bells rang, church, chapel, school, and so forth. Damage slight in poorly built buildings. Fall of plaster, in small amounts. Cracked plaster somewhat, especially fine cracks in chimneys in some instances. Broke dishes, glassware, in considerable quantity, also some windows. Fall of knick-knacks, books, pictures. Overturned furniture in many instances. Moved furnishings of moderately heavy kind.

VII

Frightened all, general alarm, all ran outdoors. Some or many found it difficult to stand. Noticed by persons driving motor cars. Trees, bushes shaken moderately to strongly. Waves on ponds, lakes, running water. Water turbid from mud stirred up. Incaving to some extent of sand or gravel stream banks. Rang large church bells, and so forth. Suspended objects made to quiver. Damage negligible of good design and construction, slight to moderate in well-built ordinary buildings, considerable in poorly built or badly designed buildings, adobe houses, old walls (especially where laid up without mortar), spires and so forth. Cracked chimneys to considerable extent, walls to some extent. Fall of plaster in considerable to large amounts, also some stucco. Broke numerous windows, furniture, to some extent. Shook down loosened brickwork and tiles. Broke weak chimneys at the roof line, sometimes damaging roofs. Fall of cornices from towers, high buildings. Dislodged bricks and stones. Overturned heavy furniture, with damage from breaking. Damage considerable to concrete irrigation ditches.

VIII

Fright general, alarm approaches panic. Disturbed persons driving motor cars. Trees shaken strongly, branches, trunks, broken off, especially palm trees. Ejected sand, mud, in small amounts. Changes: temporary, permanent; in flow of springs and wells; dry wells renewed flow; in temperature of spring and well waters. Damage slight in structures (brick) built especially to withstand earthquakes. Considerable in ordinary substantial buildings, partial collapse; racked, tumbled down, wooden houses in some cases; threw out panel walls in frame structures, broke off decayed piling. Fall of walls. Cracked, broke, solid stone walls seriously. Wet ground to some extent, also ground on steep slopes. Twisting, fall, of chimneys, columns, monuments, also factory stacks, towers. Moved conspicuously, overturned, very heavy furniture.

IX

Panic general. Cracked ground conspicuously in (masonry) structures built especially to withstand earthquakes; threw out of plumb some wood-frame houses built especially to withstand earthquakes; great in substantial (masonry) buildings, some collapse in large part; or wholly shifted frame buildings off foundations, racked frames; serious to reservoirs; underground pipes sometimes broken.

X

Cracked ground, especially where loose and wet, up to widths of several inches; fissures up to a yard in width ran parallel to canal and stream banks. Landslides considerable from river banks and steep coasts. Shifted sand and mud horizontally on beaches and flat land. Changed level of water in wells. Threw water on banks of canals, lakes, rivers, and so forth. Damage serious to dams, dikes, embankments. Severe to well-built wooden structures, bridges, some destroyed. Developed dangerous cracks in excellent brick walls. Destroyed most masonry, frame structures, also their foundations. Bent railroad rails slightly. Tore apart, crushed endwise, pipe lines buried in earth. Open cracks and broad wavy fold in cement pavements and asphalt road surfaces.

XI

Disturbances in ground many, widespread, varying with ground material. Broad fissures, earth slumps, land slips in soft, wet ground. Ejected water in large amounts charged with sand and mud. Caused sea waves ("tidal" waves) of significant magnitude. Damage severe to wood-frame structures, especially near shock centers. Great to dams, dikes, embankments, often for long distances. Few, if any, (masonry) structures remained standing. Destroyed large well-built bridges by the wrecking of supporting piers, pillars. Affected yielding wooden bridges less. Bent railroad rails greatly, thrust them endwise. Put pipe lines buried in the earth completely out of service.

XII

Damage total, practically all works of construction damaged greatly or destroyed. Disturbances in ground great, varied, numerous shearing cracks. Landslides, falls of rock of significant character, slumping of river banks, etc., numerous, extensive. Wrenched loose, tore off, large rock masses. Fault slips in firm rock, with notable horizontal, vertical offset displacements. Water channels, surface, underground, disturbed and modified greatly. Dammed lakes, produced waterfalls, deflected rivers, and so forth. Waves seen on ground surfaces (actually seen, probably, in some cases). Distorted lines of sight and level. Threw objects upward into the air.

Appendix 2
Southern California Earthquake Preparedness Project

California's SCEPP is probably the most comprehensive project ever undertaken to help a region prepare for a major earthquake. The following is a brief outline of suggestions from their literature that can be adapted to help the Mississippi Valley prepare for major seismic activity. Part 1 outlines steps local governments—cities and counties—can take (see Federal Emergency Management Agency, Southern California Earthquake Preparedness Project, *Comprehensive Earthquake Preparedness Planning Guidelines: City*, Earthquake Hazards Reduction Series, no. 2 [Washington: GPO, 1985]). Part 2 outlines steps corporations and other businesses can take (see FEMA, SCEPP, *Comprehensive Earthquake Preparedness Planning Guidelines: Corporate*, Earthquake Hazards Reduction Series, no. 4 [Washington: GPO, 1985]).

1 Earthquake Preparation Steps for Local Governments

Initiating a Program

First, a local governmental body should pass a resolution stressing the importance of earthquake planning and that body's commitment to seeing that a plan is developed and implemented.

The body should next appoint a planning committee to undertake the work required to develop this plan. Members should be chosen from various government departments and agencies as well as from outside government, as needed.

The committee should then identify any related activity or programs that are currently in place and see how these can be coordinated with the overall effort.

Next, the committee should make a detailed assessment of potential damage to the city or county—an assessment of both public and private sector vulnerability.

Finally, the committee should assign committee members to the various tasks dictated by the goals and objectives defined by the committee and establish "due dates" for the various segments of the program.

Program Components

Long-Term Preparedness Functions   *Before* an earthquake strikes, governments should:

1. enact seismic safety planning by arranging for future development in the city or county taking the seismic risk into account (by land-use management, zoning ordinances, seismic building codes, and so forth);

2. offer mitigation incentives by considering, for example, tax incentives for those taking specific mitigation actions, such as building earthquake-resistant structures and retrofitting existing structures;

3. establish mutual aid and joint powers agreements, by entering into agreements with surrounding communities or counties to share information, inventories, and services that are necessary to plan for earthquakes;

4. locate disaster assistance programs by determining what kinds of assistance may be available from state and federal sources both to help with planning efforts *and* to help with disaster recovery operations;

5. offer earthquake preparedness programs and train community groups and organizations on how to be self-sufficient immediately following an earthquake and on what they can do to help others following such a disaster, by staging earthquake drills in schools, for example;

6. assess community and county hazards by taking inventory and classifying all structures that may be vulnerable to earthquakes and determining what steps can be taken to lessen these vulnerabilities;

7. distribute public information by finding the most appropriate vehicle to dispense pertinent information to citizens and residents concerning earthquake preparedness and mitigation efforts (one option might be to provide the media with updates on efforts, including available programs and information bulletins);

8. identify emergency shelter and mass care facilities by locating appropriate buildings, services, supplies, and personnel that may be needed following an earthquake and, in tandem, develop specific plans to make maximum use of these resouces immediately following an earthquake.

Emergency Response and Recovery Functions *After* an earthquake, governments should:

1. identify major transportation routes, as well as alternates to be used if the major routes are out of commission, and provide for appropriate police service in rerouting traffic;

2. arrange for *reliable* emergency communication systems to become operational following an earthquake so officials and vital service groups can communicate with one another;

3. arrange for crews to begin removing, hauling, and disposing of debris that hampers emergency operations and to continue the removal of remaining debris following the emergency period;

4. be sure firefighters are employing proper disaster responses, especially as they relate to containing and neutralizing hazardous chemical spills;

5. employ trained personnel to make early assessments of buildings and other structures in the community or county to determine if they are safe to re-enter, or if they should be condemned and demolished;

6. develop a plan to coordinate all vital emergency and recovery services immediately following the earthquake so that services are available where needed most and in the shortest amount of time with the least confusion (this involves coordinating volunteer groups, distributing consumables—such as food, clothing, bedding, water, and energy—and providing emergency shelter);

7. appoint an official to provide necessary information, including updates, to the media so citizens and residents are fully informed on recovery operations and on where needed goods and services are available;

8. make sure crews are available to restore water, electricity, and gas, if such services have been damaged;

9. mobilize and coordinate medical personnel following the earthquake and designate treatment centers for their use;

10. provide for long-term recovery of the city or county through a series of redevelopment plans (concerning reconstruction and sources of available capital, for example).

2 Earthquake Preparation Steps for Corporations

Initiating a Program

Ideally, businesses should tie their earthquake plans to those of local government, but it is also necessary for them to have separate plans unique to the facilities. Local governments will have their hands full with their own problems following a major earthquake and will not have the capabilities to respond to the needs of business.

Businesses should first identify their principal areas of concern (asset and employee vulnerability, for example) and determine their levels of managerial commitment and their available resources.

A coordinating committee should be formed to develop a detailed work program and schedule, which will serve as a management tool to establish and meet deadlines and allocate resources.

Next, the committee should assess *specific* impacts of an earthquake on the operation: lifeline disruption, loss of production, effect on marketplace, loss of computer facilities, and so forth. The committee should then review what plans (if any) are already in existence and see how they can or should be improved. Finally, once the complete plan is ready, management workshops and operational drills should be scheduled to familiarize all levels of the company (management and employees) with it.

Program Components

Long-Term Preparedness Functions Before an earthquake strikes, businesses should:

1. take action to reduce risk to employees and assets by identifying structural hazards and take steps to mitigate or eliminate these hazards;

2. take the same action as in the first step with nonstructural items such as machinery, computer systems, and inventory;

3. inform employees about the earthquake risk, possible scenarios, and steps they should take to protect themselves on the job (and themselves and their families off the job) during and immediately after an earthquake;

4. develop mutual aid plans with neighboring businesses/industries to assist each other following an earthquake if one establishment happens to be severely damaged while the other is in a position to provide assistance;

5. inform customers or clients about the earthquake potential and how an earthquake would affect the business's ability to continue to serve them (such as what alternative plans are in place and where and how customers can take advantage of these alternative plans);

6. identify a solid structure that can serve as a temporary shelter and emergency services station and outfit it appropriately;

7. make sure the plan has provisions for any other steps managemment will need to take to deal with the disaster in the hours and days following.

Emergency Response and Recovery Functions *After* an earthquake, businesses should:

1. conduct a survey of the facilities to determine such things as the extent of damage, how safe the structures are, and the viability of continuing operations;

2. share information with employees, their families, corporate headquarters, the media, emergency service organizations, and others;

3. direct emergency operations and protect company personnel and assets;

4. be sure emergency medical care is available for employees and visitors;

5. arrange to have order restored and the facilities cleaned up after the emergency period in order to determine when and how business operations can be restored;

6. following clean-up operations, determine how best to resume operations with the facilities and equipment still functional and arrange for the replacement/repair of equipment and facilities not currently functional;

7. relay information to customers and clients about the business's status (in addition, retailers may wish to provide detailed information to the community in general—hardware stores on how to make repairs, banks on how to apply for loans, and so on);

8. contact other businesses and local government officials to discuss needs *and* ways to assist one another in recovery operations.

Appendix 3
Safety Instructions

The following is a condensed and combined version of material found in "Earthquake Safety and Survival Rules," *Earthquake Information Bulletin*, May–June 1975, 19–20, and *In Time of Emergency: A Citizen's Handbook* (Washington: GPO, 1985).

During an Earthquake

When you feel the rumbling or the initial shock

- *If indoors*, stay indoors. *Never* run outside.
- Take cover under a heavy, sturdy desk, table, or bench.
- If possible, pull a mattress or other thick, soft object over your head and body.
- If no protection is available, drop to the floor, curl up in a ball, and cover your head with your hands.
- Stay clear of: doorways, walls, windows, tall furniture, and basements.
- If you're in a high-rise, stay away from elevators and stairwells.
- *If outdoors*, run to an open area, away from buildings, utility poles, and other hazards. *Never* run inside.
- If in an automobile, pull over to the side of the road as quickly as is safely possible and stop. *Never* stop on a bridge.

After an Earthquake

• Check for injuries to yourself and others. Administer first aid if appropriate.

• Don't light matches, candles, or lighters until you are sure there is no leaking gas.

• Check for gas leaks, water leaks, and electrical circuit damage. Shut off gas, water, and/or electricity if appropriate (if any of the systems are damaged and/or leaking).

• Keep children safe and relaxed.

• Go through the house *carefully* to check for damage. Survey the outside for such dangers as loose chimneys.

• Leave homes or buildings that have been damaged until a safety assessment can be made.

• Don't drink the water until you're sure it's safe, particularly if it looks dirty.

• Use the telephone only in an emergency.

• Listen to your portable radio for more information.

• Be very cautious about further activities, since aftershocks may occur, or since the initial shock may have been a foreshock to a stronger shock.

Notes

Introduction

1. House Committee on Science and Technology. Subcommittee on Investigations and Oversight and Subcommittee on Science, Research, and Technology. *Earthquakes in the Eastern United States: Hearings.* 98th Cong., 2d sess., 1984, H. Doc. 141, 11.

1 The Destruction of the Mississippi Valley

1. For accounts of the 1811–1812 earthquakes and the events just prior to them, I have relied on a number of primary and secondary sources. Among the most helpful, though, have been Margaret Ross, "The New Madrid Earthquake," *Arkansas Historical Quarterly* 27 (Summer 1968): 83–104; F. A. Sampson, "The New Madrid and Other Earthquakes of Missouri," *Missouri Historical Review,* July 1913, 179–99; Wayne John Viitanen, "The Winter the Mississippi Ran Backwards: The Impact of the New Madrid, Missouri, Earthquake of 1811–1812 on Life and Letters in the Mississippi Valley" (Ph.D. diss., Southern Illinois University at Carbondale, 1972); Myron L. Fuller, *The New Madrid Earthquake,* United States Geological Survey Bulletin no. 494 (Washington: GPO, 1912); Myron L. Fuller, "Our Greatest Earthquakes," *Popular Science Monthly,* July 1906, 76–86; Robert Sidney Douglass, *History of Southeast Missouri: A Narrative Account of Its Historical Progress, Its People, and Its Principal Interests* (Chicago: Lewis, 1912); James Lal Penick, Jr., *The New Madrid Earthquakes,* rev. ed. (Columbia: Univ. of Missouri Press, 1981); Timothy Flint, *Recollections of the Last Ten Years in the Valley of the Mississippi* (Boston: Cummings, Hilliard, 1826; reprint, New York: Da Capo Press, 1968); and J. H. Tatsch, *Earthquakes: Cause, Prediction, and Control* (Sudbury, Mass.: Tatsch Associates, 1977).

2. Lynn Morrow, "New Madrid and Its Hinterland: 1783–1826," *Missouri Historical Society Bulletin,* July 1980, 246.

3. Viitanen, "The Winter the Mississippi Ran Backwards," 170–71.

4. Viitanen, "The Winter the Mississippi Ran Backwards," 170.

5. Ross, "The New Madrid Earthquake," 94.

6. Douglass, *History of Southeast Missouri,* 214.

7. Garland C. Broadhead, "The New Madrid Earthquake," *American Geologist* 30 (Aug. 1902): 76.

8. David D. March, *The History of Missouri* (New York: Lewis Historical Publishing, 1967), 271.

9. Douglass, *History of Southeast Missouri,* 227.

10. Ross, "The New Madrid Earthquake," 96.

11. Fuller, *The New Madrid Earthquake,* 91.

12. An excellent account of the destruction caused by the earthquakes throughout the Eastern United States can be found in Samuel L. Mitchill, "A Detailed Narrative of the Earthquakes which occurred on the 16th day of December, 1811. . . ." *Transactions of the Literary and Philosophical Society of New York,* 1815, 281–307. I have relied heavily on this source in this section.

13. Lewis Collins, *History of Kentucky,* vol. 2 (Covington, Ky.: Collins, 1882), 283.

14. Jewell Mayes, "Missouri History Not Found in Textbooks," *Missouri Historical Review,* Oct. 1927, 270.

15. Mitchill, "A Detailed Narrative of the Earthquakes," 287.

16. Mitchill, "A Detailed Narrative of the Earthquakes," 291–92.

17. Mitchill, "A Detailed Narrative of the Earthquakes," 282–83.

18. Mitchill, "A Detailed Narrative of the Earthquakes," 284.

19. See Wilbur A. Nelson, "Reelfoot: An Earthquake Lake," *National Geographic,* Jan. 1924, 105.

20. Flint, *Recollections of the Last Ten Years,* 226.

21. Ross, "The New Madrid Earthquake," 97–98.

22. Otto W. Nuttli, "The Mississippi Valley Earthquakes of 1811 and 1812: Intensities, Ground Motion, and Magnitudes," *Bulletin of the Seismological Society of America,* 63 (Feb. 1973): 230; and Reginald Aldworth Daly, *Our Mobile Earth* (New York: Scribner's, 1926), 22.

23. Fuller, *The New Madrid Earthquake,* 98–99.

24. Flint, *Recollections of the Last Ten Years,* 225.

25. Fuller, *The New Madrid Earthquake,* 48–59.

26. Penick, *The New Madrid Earthquakes,* 90.

27. William Atkinson, "Whole Lotta Shakin' Goin' On," *St. Louis Magazine,* Aug. 1982, 18; and "Magnitude and Intensity: Measures of Earthquake Size and Severity," *Earthquake Information Bulletin,* Sept.-Oct. 1974, 24.

28. Mayes, "Missouri History," 270.

29. For such reasons as those stated in the text, Audubon's account, although rather lengthy and interesting, has not been depended on for this study. Audubon's account can be found in Maria R. Audubon, *Audubon and His Journals*, vol. 2 (New York: Scribner's, 1897; reprint, New York: Dover Publications, 1960), 234–37; and Alice Ford, ed., *Audubon, by Himself: A Profile of John James Audubon* (Garden City, N.Y.: Natural History Press, 1969), 55–57.

2 Centuries of Mississippi Valley Earthquakes

1. Carl A. von Hake, "Earthquake History of Missouri," *Earthquake Information Bulletin*, May-June 1974, 25; and Louis Winkler, "Catalog of U.S. Earthquakes Before the Year 1850," *Bulletin of the Seismological Society of America* 69 (Apr. 1979): 598.

2. C. Simon, "Arkansas: Shakes, Rattles, but No Roll," *Science News*, 1 Jan. 1983, 7.

3. For information relating to moderate and major earthquakes since the 1811–1812 sequence, I have relied on a number of sources. The most helpful, however, have been Otto W. Nuttli, "Damaging Earthquakes of the Central Mississippi Valley," in *Investigations of the New Madrid, Missouri, Earthquake Region*, Geological Survey Professional Paper no. 1236, ed. F. A. McKeown and L. C. Pakiser (Washington: GPO, 1982), 15–20; Jerry L. Coffman, Carl A. von Hake, and Carl W. Stover, *Earthquake History of the United States*, rev. ed., prepared for the U.S. Department of Commerce (Washington: GPO, 1982): 46–57; and Margaret G. Hopper, S. T. Algermissen, and Ernest E. Dobrovolny, *Estimation of Earthquake Effects Associated with a Great Earthquake in the New Madrid Seismic Zone*, U.S. Geological Survey Open-File Report no. 83–179 (Washington: GPO, 1983). Other helpful sources have been issues of the *St. Louis Globe-Democrat, St. Louis Post-Dispatch, Chicago Daily Tribune, Chicago Sun-Times, Memphis Commercial Appeal, Cairo Evening Citizen*, and *Southern Illinoisan*.

3 Earthquake Theory

1. "Chinese Earthquake: Latest in a Chain That Raises Worldwide Concern," *U.S. News and World Report*, 9 Aug. 1976, 18.

2. "The Next Big Earthquake Is on Its Way," *Changing Times*, Feb. 1977, 37.

3. "Archivist Disputes 1906 S. F. Death Toll," *USA Today*, 18 Apr. 1985, sec. A, 6.

4. James Cornell, *The Great International Disaster Book* (New York: Pocket Books, 1979), 144.

5. Richard L. Williams, "Science Tries To Break New Ground in Predicting Earthquakes," *Smithsonian*, July 1983, 42.

6. "Quake Plan Asks $366 Million," *Engineering News Record*, 17 Jan. 1985, 15.

7. "Recent Quakes Shake Our Awareness," *Southern Illinoisan*, 23 Oct. 1985, 22.

8. "Scientists Closing In on Ability To Pinpoint Event," *Science Digest*, Mar. 1978, 14; William J. Petak and Arthur A. Atkisson, *Natural Hazard Risk Assessment and Public Policy: Anticipating the Unexpected* (New York: Springer-Verlag, 1982), 27–28.

9. For basic reading on plate tectonics, see Henry Spall, "Earthquakes and Plate Tectonics," *Earthquake Information Bulletin*, Nov.-Dec. 1977, 14–17; Henry Spall, "Paleomagnetism: One Key to Plate Tectonics," *Earthquake Information Bulletin*, Jan.-Feb. 1973, 13–17; and Rick Gore, "Our Restless Planet Earth," *National Geographic*, Aug. 1985, 142–72.

10. Ben-chieh Liu et al., *Earthquake Risk and Damage Functions: Application to New Madrid* (Boulder, Colo.: Westview Press, 1981), 8.

11. For my discussion of the genesis of the New Madrid Seismic Zone, I have relied primarily on the following sources: Robert M. Hamilton, "Quakes Along the Mississippi," *Natural History*, Aug. 1980, 70–75; Arch C. Johnston, "A Major Earthquake Zone on the Mississippi," *Scientific American*, Apr. 1982, 60–68; M. F. Kane, T. G. Hildenbrand, and J. D. Hendricks, "Model for the Tectonic Evolution of the Mississippi Embayment and Its Contemporary Seismicity," *Geology* 9 (Dec. 1981): 563–68; Richard A. Kerr, "Assessing the Risk of Eastern U.S. Earthquakes," *Science*, 9 Oct. 1981, 169–71; Lawrence Malinconico, interview with author, Carbondale, Ill., 3 Dec. 1985; and Frank A. McKeown, telephone interview with author, 5 Dec. 1985.

12. House Committee on Science and Technology, *Earthquakes in the Eastern United States*, 90; and Nuttli, "The Mississippi Valley Earthquakes of 1811 and 1812," 247.

13. These two theories are discussed in more detail in Mark D. Zoback and Mary Lou Zoback, "State of Stress and Intraplate Earthquakes in the United States," *Science*, 3 July 1981, 101–2; Kerr, "Assessing the Risk," 169–70; and Hamilton, "Quakes Along the Mississippi," 75.

14. This theory has been proposed a number of times in the past. For a discussion of the theory, past and present, see Lawrence W. Braile et al., "The Northeastern Extension of the New Madrid Seismic Zone," in *Investigations of the New Madrid, Missouri, Earthquake Region*, Geological Survey Professional Paper no. 1236, ed. F. A. McKeown and L. C. Pakiser (Washington: GPO, 1982), 175, 183–84. Other authorities, however, among them Otto W. Nuttli (telephone interview with the author) and Lawrence Malinconico (interview with the author), state that there is no evidence of such a link.

4 Earthquake Prediction

1. Cornell, *The Great International Disaster Book*, 36; and Peter L. Ward, "Earthquake Prediction: A Perspective for a National Program," *Earthquake Information Bulletin*, Nov.-Dec. 1977, 10.

2. Cornell, *The Great International Disaster Book*, 36; and "Chinese Earthquake," 22.

3. Ward, "Earthquake Prediction," 10.

4. "Forecast: Future Shock," *Time*, 24 Jan. 1977, 83.

5. "Forecasting of Earthquakes Shows Gains," *U.S. News and World Report*, 30 Apr. 1979, 65.

6. "Prospects for Earthquake Prediction Wane," *Science*, 2 Nov. 1979, 543.

7. Ronald Schiller, "Where Will the Next Earthquake Strike?" *Reader's Digest*, Oct. 1979, 124.

8. Allen Lindh, "Amplitude of Foreshocks as a Possible Precursor to Earthquakes," *Earthquake Information Bulletin*, Sept.-Oct. 1978, 162.

9. Helmut Tributsch, *When the Snakes Awake: Animals and Earthquake Prediction*, trans. Paul Langner (Cambridge, Mass.: MIT Press, 1982), 89–91.

10. Frank Press, "Earthquake Prediction," *Scientific American* 232 (May 1975): 17.

11. Chi-Yu King, "Electromagnetic Emissions Before Earthquakes," *Nature* 301 (Feb. 1983): 377.

12. John S. Derr, "Earthquake Lights: A Review of Observations and Present Theories," *Bulletin of the Seismological Society of America* 63 (Dec. 1973): 2178.

13. Derr, "Earthquake Lights: A Review," 2179.

14. William R. Corliss, *Earthquakes, Tides, Unidentified Sounds, and Related Phenomena: A Catalog of Geophysical Anomalies* (Glen Arm, Md.: Sourcebook Project, 1983), 83–84.

15. Frank A. McKeown, telephone interview with author, 5 Dec. 1985.

16. G. Lennis Berlin, *Earthquakes and the Urban Environment*, vol. 2 (Boca Raton, Fla.: CRC Press, 1980), 59.

17. Berlin, *Earthquakes and the Urban Environment* 2:62.

18. Berlin, *Earthquakes and the Urban Environment* 2:62–63.

19. See Roger N. Hunter, "Is There a Periodicity in the Occurrence of Earthquakes?" *Earthquake Information Bulletin*, Sept.-Oct. 1976, 5; and Corliss, *Earthquakes, Tides, Unidentified Sounds, and Related Phenomena*, 97. The author highly recommends Corliss's book for readers interested in detailed discussion of earthquake periodicity.

20. Corliss, *Earthquakes, Tides, Unidentified Sounds, and Related Phenomena*, 108–9; and Richard F. Haines, ed., *UFO Phenomena and*

the Behavioral Scientist (Metuchen, N.J.: Scarecrow Press, 1979), 396–431.

21. Corliss, *Earthquakes, Tides, Unidentified Sounds, and Related Phenomena,* 108–9.

22. John Milne, *Earthquakes and Other Earth Movements* (New York: Appleton, 1886), 263.

23. See Hunter, "Is There a Periodicity in the Occurrence of Earthquakes?" 5; Patrick Huyghe, "Earthquakes: The Solar Connection," *Science Digest,* Oct. 1982, 74; "Earthquakes May Be Triggered by the Sun," *New Scientist,* 16 May 1968, 357; and Corliss, *Earthquakes, Tides, Unidentified Sounds, and Related Phenomena,* 98–102.

24. See Corliss, *Earthquakes, Tides, Unidentified Sounds, and Related Phenomena,* 120–27; "Earthquakes May Parallel, Not Cause, Polar Wobble," *New Scientist and Science Journal,* 21 Jan. 1971, 105; "Earthquakes and the Rotation of the Earth," *Science,* 4 Oct. 1974, 49–50; Frank Press and Peter Briggs, "Chandler Wobble, Earthquakes, Rotation, and Geomagnetic Changes," *Nature* 256 (24 July 1975): 270–73; Hiroo Kanamori, "Are Earthquakes a Major Cause of the Chandler Wobble?" *Nature* 262 (22 July 1976): 254–55; and Richard J. O'Connell and Adam M. Dziewonski, "Excitation of the Chandler Wobble by Large Earthquakes," *Nature* 262 (22 July 1976): 259–62.

25. Milne, *Earthquakes and Other Earth Movements,* 251; "Periodicity in Earthquakes," *Nature,* 20 Oct. 1934, 631; "Sun and Moon Tidal Forces May Trigger Some Classes of Earthquakes," *Earthquake Information Bulletin,* May-June 1975, 15; and Corliss, *Earthquakes, Tides, Unidentified Sounds, and Related Phenomena,* 103–7.

26. Milne, *Earthquakes and Other Earth Movements,* 256–58.

27. Clarence R. Allen, "Earthquake Prediction: 1982 Overview," *Bulletin of the Seismological Society of America* 72 (Dec. 1982): S333; Gurney Williams III, "How Science Forecasts the Killer Quakes," *Popular Mechanics,* Aug. 1981, 105–6; Otto W. Nuttli, telephone interview with author, 4 Dec. 1985; and Lawrence Malinconico, interview with author, Carbondale, Ill., 3 Dec. 1985.

28. For excellent overviews of animal earthquake prediction, the author recommends Tributsch, *When the Snakes Awake;* and Corliss, *Earthquakes, Tides, Unidentified Sounds, and Related Phenomena.* Other sources used for this section include: David Monagan, "How Animals Predict Earthquakes," *Science Digest,* June 1981, 92–95, 124; Phylis Magida, "If Pandas Scream . . . an Earthquake Is Coming!" *International Wildlife,* Sept. 1977, 37–39; Evelyn Shaw, "Can Animals Anticipate Earthquakes?" *Natural History,* Nov. 1977, 14–20; Shen Ling-huang, "Can Animals Help to Predict Earthquakes?" *Earthquake Information Bulletin,* Nov.-Dec. 1978, 231–33; Paul Reasenberg, "Unusual Animal Behavior Before Earthquakes," *Earthquake Information Bulletin,* Jan.-Feb. 1978,

42–49; and Ruth B. Simon, "Animal Behavior and Earthquakes," *Earthquake Information Bulletin*, Nov.-Dec. 1975, 9–11.

29. Shaw, "Can Animals Predict Earthquakes?" 14.

30. Shaw, "Can Animals Predict Earthquakes?" 14.

31. Ling-huang, "Can Animals Help to Predict Earthquakes?" 232.

32. Tsuneji Rikitake, "Anomalous Animal Behavior Preceding the 1978 Earthquake of Magnitude 7.0 That Occurred near Izu-Oshima Island, Japan," in *Current Research in Earthquake Prediction*, vol. 1, ed. Tsuneji Rikitake (Dordrecht, Holland: Reidel, 1981), 75–80.

33. Tributsch, *When the Snakes Awake*, 170–71.

34. Richard A. Kerr, "Quake Prediction by Animals Gaining Respect," *Science*, 16 May 1980, 695.

35. Huyghe, "Earthquakes: The Solar Connection," 73–74.

36. Otto W. Nuttli, telephone interview with author, 4 Dec. 1985.

37. The best known study of radon emanation from the New Madrid Seismic Zone is S. R. Steele, W. C. Hood, and J. L. Sexton, "Radon Emanation in the New Madrid Seismic Zone," in *Investigations of the New Madrid, Missouri, Earthquake Region*, Geological Survey Professional Paper no. 1236, ed. F. A. McKeown and L. C. Pakiser (Washington: GPO, 1982), 191–201. Other noted experts, however, dispute the value of radon emanation as an accurate earthquake precursor in the Midwest: Lawrence Malinconico, interview with author, Carbondale, Ill., 3 Dec. 1985; and Frank A. McKeown, telephone interview with author, 5 Dec. 1985.

38. See "Assessing the Risk of Earthquakes in the Eastern United States," *Earthquake Information Bulletin*, Sept.-Oct. 1981, 177.

39. "Assessing the Risk of Earthquakes," 177.

40. A good overview of Midwest seismicity with a special emphasis on frequency estimates is F. A. McKeown, D. P. Russ, and P. C. Thenhaus, "Central Interior Seismic Source Zones: Summary of Workshop Convened June 10–11, 1980," in *Summary of Workshops Concerning Regional Seismic Source Zones as Parts of the Conterminous United States, Convened by the Geological Survey 1979–1980, Golden, Colo.*, Geological Survey Circular no. 898, ed. Paul C. Thenhaus (Washington: GPO, 1983), 19–24.

41. House Committee on Science and Technology, *Earthquakes in the Eastern United States*, 11, 143.

42. Otto W. Nuttli, telephone interview with author, 4 Dec. 1985.

43. Ward, "Earthquake Prediction," 11.

44. Frederick Ahearn, telephone interview with author, 6 Dec. 1985.

45. Berlin, *Earthquakes and the Urban Environment* 3:81.

46. Dennis Mileti, telephone interview with author, 20 Dec. 1985.

47. Janice R. Hutton, John H. Sorensen, and Dennis S. Mileti, "Earth-

quake Prediction and Public Reaction," in *Current Research in Earthquake Prediction* 1:147.

48. "Forecast: Future Shock," 83.

49. C. B. Raleigh, "Scientists' Responsibility for Public Information," *Science*, 31 July 1981, 499.

5 Earthquake Preparedness

1. John M. Leighty, "System 'Floats' Buildings Through Quakes," *Chicago Tribune*, 16 Sept. 1984, sec. 6, 3.

2. "Are U.S. Cities Ready for a Major Quake?" *U.S. News and World Report*, 3 Mar. 1980, 40.

3. Paul Flores, telephone interview with author, 4 Dec. 1985.

4. All quotations from Mr. Flores in this chapter, as well as all ideas attributed to him, are taken from his telephone interview with the author, 4 Dec. 1985.

5. All quotations from Dr. Moy in this chapter, as well as all ideas attributed to him, are taken from his interview with the author, 4 Dec. 1985.

6. J. James Eidel, telephone interview with author, 20 Dec. 1985.

7. Risa I. Palm, "Public Response to Earthquake Hazard Information," *Annals of the Association of American Geographers* 71 (Sept. 1981): 397.

8. Hans Fischer, interview with author, Carbondale, Ill., 13 Dec. 1985.

9. Ann Schottman Knol, "Are You Ready? Most Folks Aren't in a Rush To Prepare for an Earthquake," *Southern Illinoisan*, 28 June 1984, 1.

10. House Committee on Science and Technology, *Earthquakes in the Eastern United States*, 47.

11. Lawrence Malinconico, interview with author, Carbondale, Ill., 3 Dec. 1985.

12. Frederick Ahearn, telephone interview with author, 6 Dec. 1985.

13. Anonymous source, confidential telephone interview with author, 6 Dec. 1985.

14. Joanne Nigg quoted in House Committee on Science and Technology, *Earthquakes in the Eastern United States*, 52.

15. All quotations from Dr. Quarantelli, as well as all ideas attributed to him, are taken from his telephone interview with the author, 10 Dec. 1985.

16. Dennis Mileti, telephone interview with author, 20 Dec. 1985.

17. "Is a Killer Earthquake Due to Hit the U.S.?" *U.S. News and World Report*, 14 Nov. 1983, 76.

18. "Earthquake: Scientists Say Missouri Bootheel Is Likely Center of Next Big One," *Southern Illinoisan*, 25 May 1982, 5.

6 The Next Twenty Years

1. All quotations from Dr. Mileti in this chapter, as well as all ideas attributed to him, are taken from his telephone interview with the author, 20 Dec. 1985.

2. K. Jill Kiecolt and Joanne M. Nigg, "Mobility and Perceptions of a Hazardous Environment," *Environment and Behavior* 14 (March 1982): 152.

3. All quotations from Mr. Fischer in this chapter, as well as all ideas attributed to him, are taken from his interview with the author, Carbondale, Ill., 13 Dec. 1985.

4. All quotations from Dr. Schiff in this chapter, as well as all ideas attributed to him, are taken from his telephone interview with the author, 4 Dec. 1985.

5. All quotations from Dr. Johnston, as well as all ideas attributed to him, are taken from his telephone interview with the author, 6 Dec. 1985.

6. Anonymous source, telephone interview with author, 6 Dec. 1985.

7. All quotations from Mr. Flores in this chapter, as well as all ideas attributed to him, are taken from his telephone interview with the author, 4 Dec. 1985.

8. All quotations from Dr. Moy in this chapter, as well as all ideas attributed to him, are taken from his interview with the author, Carbondale, Ill., 11 Dec. 1985.

9. An excellent source of information on this subject is Raquel E. Cohen, M.D., and Frederick L. Ahearn, Jr., *Handbook for Mental Health Care of Disaster Victims* (Baltimore: Johns Hopkins Univ. Press, 1980).

10. All quotations from Dr. Ahearn in this chapter, as well as all ideas attributed to him, are taken from his telephone interview with the author, 6 Dec. 1985.

11. All quotations from Dr. Malinconico in this chapter, as well as all ideas attributed to him, are taken from his interview with the author, Carbondale, Ill., 3 Dec. 1985.

12. All quotations from Dr. Quarantelli in this chapter, as well as all ideas attributed to him, are taken from his telephone interview with the author, 10 Dec. 1985.

13. Wilbur R. Buntin, Jr., *Report to Governor Martha Layne Collins,* prepared for the Governor's Earthquake Hazards and Safety Technical Advisory Panel (Frankfort, Ky.: Division of Disaster and Emergency Services, 1985), 5.

14. Howard Kunreuther et al., *Disaster Insurance Protection: Public Policy Lessons* (New York: Wiley, 1978), 7.

15. For detailed information on building homes designed to survive

earthquakes, see Federal Emergency Management Agency, *The Home Builder's Guide for Earthquake Design* (Washington: GPO, 1985). I am thankful to Hans Fischer for information on home construction.

16. "Childhood Pressures," *USA Today,* 27 June 1983.

7 Surviving the Next Major New Madrid Earthquake

1. Liu et al., *Earthquake Risk and Damage Functions,* 183.

2. Hopper, Algermissen, and Dobrovolny, *Estimation of Earthquake Effects,* vi; and Allen and Hoshall, Inc., *An Assessment of Damage and Casualties for Six Cities in the Central United States Resulting from Earthquakes in the New Madrid Seismic Zone,* prepared for the Federal Emergency Management Agency (Washington: GPO, 1985), i.

3. David Alexander, "Death and Injury in Earthquakes," *Disasters,* 1 Sept. 1985, 58.

4. These maps are based on a map presented in Central United States Earthquake Consortium, *The First Year: 1985,* (n.p., 1985).

5. All quotations from Dr. Johnston and Dr. Nuttli in this chapter, as well as all ideas attributed to them, are taken from their telephone interviews with the author, 6 Dec. 1985 and 4 Dec. 1985 respectively.

6. All quotations from Dr. Malinconico in this chapter, as well as all ideas attributed to him, are taken from his interview with the author, Carbondale, Ill., 3 Dec. 1985.

7. Allen and Hoshall, *An Assessment of Damage and Casualties,* v, xi.

8. Allen and Hoshall, *An Assessment of Damage and Casualties,* v, xi.

9. All information for this section ("Medical Needs") is the result of Dr. Moy's interview with the author, Carbondale, Ill., 11 Dec., 1985.

10. All quotations from Dr. Mileti in this chapter, as well as all ideas attributed to him, are taken from his telephone interview with the author, 20 Dec. 1985.

11. All quotations from Dr. Quarantelli in this chapter, as well as all ideas attributed to him, are taken from his telephone interview with the author, 10 Dec. 1985.

12. All quotations from Dr. Ahearn in this chapter, as well as all ideas attributed to him, are taken from his telephone interview with the author, 6 Dec. 1985.

13. All quotations from Mr. Ertle in this chapter, as well as all ideas attributed to him, are taken from his telephone interview with the author, 1 May 1984.

14. G. Barbina, "The Fruili Earthquake as an Agent of Social Change in a Rural Area," *Mass Emergencies* 4 (1979): 145–49.

15. As sources for my discussion of the Buffalo Creek disaster, I have used primarily K. T. Erickson, "Loss of Communality at Buffalo Creek," *American Journal of Psychiatry* 133 (1976): 302–5; L. Rangell, "Discussion of the Buffalo Creek Disaster: The Course of Psychic Trauma," *American Journal of Psychiatry* 133 (1976): 313–16; Robert Jay Lifton and Eric Olson, "The Human Meaning of Total Disaster: The Buffalo Creek Experience," *Psychiatry* 39 (Feb. 1976); 1–18: and G. C. Gleser, B. L. Green, and C. C. Winget, *Prolonged Psychosocial Effects of Disaster: A Study of Buffalo Creek* (New York: Academic Press, 1981).

16. Cohen and Ahearn, *Handbook for Mental Health Care,* 30–31.

17. Glin Bennet, *Beyond Endurance: Survival at the Extremes,* (New York: St. Martin's Press, 1983), 175–76, 179.

18. S. T. Boyd, "Psychological Reactions of Disaster Victims," *South African Medical Journal* 60 (1981): 744–48; and J. G. Edwards, "Psychiatric Aspects of Civilian Disasters," *British Medical Journal* 1 (1976): 944–47.

19. Michael A. Crabbs and Edward Heffron, "Loss Associated with a Natural Disaster," *Personnel and Guidance Journal,* Feb. 1981, 379.

20. R. R. Koegler and S. M. Hicks, "The Destruction of a Medical Center by Earthquake: Initial Effects on Patients and Staff," *California Medicine* 116 (1972): 63–67.

21. Frederick L. Ahearn, Jr., "Disaster Mental Health: A Pre- and Post-Earthquake Comparison of Psychiatric Admission Rates," *Urban and Social Change Review,* Winter 1981, 27.

22. Information for this section ("The Children") is from the following sources: Berlin, *Earthquakes and the Urban Environment* 3:94–95; Crabbs and Heffron, "Loss Associated with a Natural Disaster," 380; C. J. Newman, "Children of Disaster: Clinical Observations at Buffalo Creek," *American Journal of Psychiatry* 133 (1976): 306–12; and Ahearn telephone interview.

Bibliography

The following bibliography is divided into sections listing books and pamphlets, articles, public documents, unpublished material, newspapers, and interviews. Sources marked with an asterisk are recommended reading.

Books and Pamphlets

Adams, William Mansfield. *Earthquakes: An Introduction to Observational Seismology.* Boston: Heath, 1964.

American Red Cross. *Safety and Survival in an Earthquake.* Los Angeles: American Red Cross, 1984.

Asada, Toshi, ed. *Earthquake Prediction Techniques: Their Application in Japan.* Tokyo: Univ. of Tokyo Press, 1982.

Attewell, P. B., and R. K. Taylor, eds. *Ground Movements and Their Effects on Structures.* Glasgow: Surrey Univ. Press, 1984.

Audubon, Maria R. *Audubon and His Journals.* Vol. 2. New York: Scribner's, 1897. Reprint. New York: Dover, 1960.

Ayre, Robert S., Dennis Mileti, and Patricia B. Trainer. *Earthquake and Tsunami Hazards in the United States: A Research Assessment.* Boulder: Univ. of Colorado, 1975.

Bareis, Charles, and James W. Porter, eds. *American Bottom Archaeology.* Urbana: Univ. of Illinois Press, 1984.

Barton, Allen H. *Communities in Disaster: A Sociological Analysis of Collective Stress Situations.* Garden City, N.Y.: Doubleday, 1969.

———. *Social Organization Under Stress: A Sociological Review of Disaster Studies.* Washington: National Academy of Sciences, 1963.

Bateman, Newton, and Paul Selby, eds. *Historical Encyclopedia of Illinois.* Chicago: Munsell, 1907.

Båth, Markus. *Introduction to Seismology.* New York: Wiley, 1973.

Beavers, James E., ed. *Earthquakes and Earthquake Engineering: Eastern United States.* 2 vols. Ann Arbor, Mich.: Ann Arbor Science Publishers, 1981.

Bibliography

Bennet, Glin. *Beyond Endurance: Survival at the Extremes.* New York: St. Martin's Press, 1983.

* Berlin, G. Lennis. *Earthquakes and the Urban Environment.* 3 vols. Boca Raton, Fla.: CRC Press, 1980.

Bolt, Bruce, ed. *Earthquakes and Volcanoes.* San Francisco: Freeman, 1980.

Bullen, K. E. *An Introduction to the Theory of Seismology.* Cambridge: Cambridge Univ. Press, 1965.

Clifton, Juanita. *Reelfoot and the New Madrid Quake.* Asheville, N.C.: Victor, 1980.

* Cohen, Raquel E., M.D., and Frederick L. Ahearn, Jr. *Handbook for Mental Health Care of Disaster Victims.* Baltimore: Johns Hopkins Univ. Press, 1980.

Collins, Lewis. *History of Kentucky.* Vol. 2. Covington, Ky.: Collins, 1882.

Commission on Sociotechnical Systems. *A Program of Studies on the Socioeconomic Effects of Earthquake Predictions.* Washington: National Academy of Sciences, 1978.

Commonwealth Edison. *Emergency Information: 1985.* Chicago: Commonwealth Edison, 1985.

Corliss, William R., ed. *Earthquakes, Tides, Unidentified Sounds, and Related Phenomena: A Catalog of Geophysical Anomalies.* Glen Arm, Md.: Sourcebook Project, 1983.

Cornell, James. *The Great International Disaster Book.* New York: Pocket Books, 1979.

Dacy, Douglas C., and Howard Kunreuther. *The Economics of Natural Disaster: Implications for Federal Policy.* New York: Free Press, 1969.

Daly, Reginald Aldworth. *Our Mobile Earth.* New York: Scribner's, 1926.

Danzig, Elliott R., Paul W. Thayer, and Lila R. Galanter. *The Effects of a Threatening Rumor on a Disaster-Stricken Community.* National Research Council Publication no. 517. Washington: National Academy of Sciences, 1958.

De Nevi, Don. *Earthquakes.* Millbrae, Calif.: Celestial Arts, 1977.

Douglass, Robert Sidney. *History of Southeast Missouri: A Narrative Account of Its Historical Progress, Its People, and Its Principal Interests.* Chicago: Lewis, 1912.

Droessler, Judith. *Craniometry and Biological Distance: Biocultural Continuity and Change in the Late-Woodland–Mississippian Interface.* Evanston, Ill.: Center for American Archaeology, Northwestern Univ., 1981.

Dynes, Russel R. *Organized Behavior in Disaster.* Lexington, Mass.: Heath, 1970.

Eiby, G. A. *Earthquakes.* New York: Van Nostrand Reinhold, 1980.

Encyclopaedia Britannica Editors. *Catastrophe! When Man Loses Control.* New York: Bantam Books, 1979.

————. *Disaster! When Nature Strikes Back.* New York: Bantam Books, 1978.

Englekirk, Robert E., Gary C. Hart, and the Concrete Masonry Association of California and Nevada. *Earthquake Design of Concrete Masonry Buildings.* 2 vols. Englewood Cliffs, N.J.: Prentice-Hall, 1982.

Flint, Timothy. *Recollections of the Last Ten Years in the Valley of the Mississippi.* Boston: Cummings, Hilliard, 1826. Reprint. New York: Da Capo Press, 1968.

Ford, Alice, ed. *Audubon, by Himself: A Profile of John James Audubon.* Garden City, N.Y.: Natural History Press, 1969.

Friesma, H. Paul, James Caporaso, Gerald Goldstein, Robert Lineberry, and Richard McCleary. *Aftermath: Communities After Natural Disasters.* Beverly Hills, Calif.: Sage, 1979.

Fritz, Charles E., and J. H. Mathewson. *Convergence Behavior in Disasters: A Problem in Social Control.* National Research Council Publication no. 476. Washington: National Academy of Sciences, 1957.

Garb, Solomon, and Evelyn Eng. *Disaster Handbook.* New York: Spring, 1969.

Gleser, G. C., B. L. Green, and C. C. Winget. *Prolonged Psychosocial Effects of Disaster: A Study of Buffalo Creek.* New York: Academic Press, 1981.

Goodspeed's History of Southeast Missouri. Cape Girardeau, Mo.: Ramfre Press, 1955.

Gould, Phillip L., and Salman H. Abu-Sitta. *Dynamic Response of Structures to Wind and Earthquake Loading.* New York: Wiley, 1980.

Haines, Richard F., ed. *UFO Phenomena and the Behavioral Scientist.* Metuchen, N.J.: Scarecrow Press, 1979.

Haywood, John. *The Natural and Aboriginal History of Tennessee: Up to the First Settlements Therein by the White People in the Year 1768.* Reprint. Kingsport, Tenn.: Hill Books, 1973.

Healy, Richard J. *Emergency and Disaster Planning.* New York: 1969.

Heck, Nicholas Hunter. *Earthquakes.* New York: Hafner, 1965.

History of Madison County, Illinois. Edwardsville, Ill.: Brink, 1882.

Hodgson, John H. *Earthquakes and Earth Structure.* Englewood Cliffs, N.J.: Prentice-Hall, 1964.

Hollis, Edward P. *Bibliography of Earthquake Engineering.* Oakland, Calif.: Earthquake Engineering Research Institute, 1971.

Houck, Louis. *A History of Missouri: From the Earliest Explorations and Settlements Until the Admission of the State into the Union.* Chicago: Donnelley, 1908.

Bibliography

Iwan, W. D., ed. *Applied Mechanics in Earthquake Engineering*. New York: American Society of Mechanical Engineers, 1974.

* Kimball, Virginia. *Earthquake Ready*. Culver City, Calif.: Peace Press, 1981.

Kunreuther, Howard. *Recovery from Natural Disasters: Insurance or Federal Aid?* Washington: American Enterprise Institute for Public Policy Research, 1973.

Kunreuther, Howard, Ralph Ginsberg, Louis Miller, Philip Sagi, Paul Slovic, Bradley Borkin, and Norman Katz. *Disaster Insurance Protection: Public Policy Lessons*. New York: Wiley, 1978.

Laube, Jerri, and Shirley A. Murphy, eds. *Perspectives on Disaster Recovery*. Norwalk, Conn.: Appleton-Century-Crofts, 1985.

Liu, Ben-chieh, Chang-tseh Hseih, Robert Gustafson, Otto Nuttli, and Richard Gentile. *Earthquake Risk and Damage Functions: Application to New Madrid*. Boulder, Colo.: Westview Press, 1981.

Lomnitz, C., and E. Rosenblueth, eds. *Seismic Risk and Engineering Decisions*. Amsterdam: Elsevier Scientific Publishing, 1976.

March, David D. *The History of Missouri*. New York: Lewis Historical Publishing, 1967.

Mileti, Dennis. *Natural Hazard Warning Systems in the United States: A Research Assessment*. Boulder: Univ. of Colorado, 1975.

Mileti, Dennis, Thomas E. Drabek, and J. Eugene Haas. *Human Stress in Extreme Environments: A Sociological Perspective*. Boulder: Univ. of Colorado, 1975.

Milne, John. *Earthquakes and Other Earth Movements*. New York: Appleton, 1886.

National Research Council. *Field Studies of Disaster Behavior: An Inventory*. Washington: National Academy of Sciences, 1961.

———. Assembly of Mathematical and Physical Sciences. Committee on Seismology. Panel on Seismograph Networks. *Global Earthquake Monitoring: Its Uses, Potentials, and Support Requirements*. Washington: National Academy of Sciences, 1977.

———. *U.S. Earthquake Observatories: Recommendations for a New National Network*. Washington: National Academy Press, 1980.

Newmark, Nathan, and Emilio Rosenblueth. *Fundamentals of Earthquake Engineering*. Englewood Cliffs, N.J.: Prentice-Hall, 1971.

Okamoto, Shunzo. *Introduction to Earthquake Engineering*. New York: Wiley, 1973.

Parad, Howard J., H. L. P. Resnick, and Libbie G. Parad. *Emergency and Disaster Management: A Mental Health Sourcebook*. Bowie, Md.: Charles Press, 1976.

Penick, James Lal. *The New Madrid Earthquakes*. Rev. ed. Columbia, Mo.: Univ. of Missouri Press, 1981.

Perry, Ronald W. *The Social Psychology of Civil Defense*. Lexington, Mass.: Heath, 1982.

Petak, William J., and Arthur A. Atkisson. *Natural Hazard Risk Assessment and Public Policy: Anticipating the Unexpected*. New York: Springer-Verlag, 1982.

Reynolds, John. *The Pioneer History of Illinois: Containing the Discovery in 1673, and the History of the Country to the Year 1818, When the State Government Was Organized*. Chicago: Fergus Printing, 1887.

Richter, Charles F. *Elementary Seismology*. San Francisco: Freeman, 1958.

Rikitake, Tsuneji. *Earthquake Forecasting and Warning*. Dordrecht, Holland: Reidel, 1982.

————. *Earthquake Prediction*. Amsterdam: Elsevier Scientific Publishing, 1976.

————, ed. *Current Research in Earthquake Prediction*. Vol.1. Dordrecht, Holland: Reidel, 1981.

Robinson, Donald. *The Face of Disaster*. New York: Doubleday, 1959.

Rossi, Peter H., James D. Wright, Eleanor Weber-Burdin, and Joseph Pereira. *Victims of the Environment: Loss from Natural Hazards in the United States, 1970–1980*. New York: Plenum Press, 1983.

Scheidegger, Adrian E. *Physical Aspects of Natural Catastrophes*. Amsterdam: Elsevier Scientific Publishing, 1975.

Simpson, David W., and Paul G. Richards, eds. *Earthquake Prediction: An International Review*. Washington: American Geophysical Union, 1981.

Smith, D. J., Jr., ed. *Lifeline Earthquake Engineering*. New York: American Society of Civil Engineers, 1981.

Swiss Reinsurance Company. *Atlas on Seismicity and Volcanism*. Bern, Switzerland: Kümmerly and Krey, 1977.

Tatsch, J. H. *Earthquakes: Cause, Prediction, and Control*. Sudbury, Mass.: Tatsch Associates, 1977.

Thwaites, Reuben Gold, ed. *Early Western Travels, 1748–1846*. Vol. 5, *[John] Bradbury's Travels in the Interior of America, 1809–1811;* vol. 9, *[James] Flint's Letters from America, 1818–1820;* vol. 13, *[Thomas] Nuttall's Travels into the Arkansa [sic] Territory, 1819;* vol. 17, *Part IV of Jame's [sic] Account of S. H. Long's Expedition, 1819–1820;* vol. 24, *Part I of [Edmund] Flagg's The Far West, 1836–1837*. Reprint. Cleveland: Clark, 1904–1906.

Tributsch, Helmut. *When the Snakes Awake: Animals and Earthquake Prediction*. Translated by Paul Langner. Cambridge, Mass.: MIT Press, 1982.

United Nations Educational, Scientific, and Cultural Organization. *The Assessment and Mitigation of Earthquakes*. Paris: UNESCO, 1978.

Bibliography

Wallace, Anthony F. C. *Human Behavior and Extreme Situations: A Survey of the Literature and Suggestions for Further Research.* Washington: National Academy of Sciences, 1956.

White, Gilbert F., ed. *Natural Hazards: Local, National, Global.* London: Oxford Univ. Press, 1974.

White, Gilbert F., and J. Eugene Haas. *Assessment of Research on Natural Hazards.* Cambridge, Mass.: MIT Press, 1975.

Wiegel, Robert L., ed. *Earthquake Engineering.* Englewood Cliffs, N.J.: Prentice-Hall, 1970.

Williams, Samuel Cole. *Beginnings of West Tennessee: In the Land of the Chickasaws, 1541–1841.* Johnson City, Tenn.: Watauga Press, 1930.

Wolfenstein, Martha. *Disaster: A Psychological Essay.* Glencoe, Ill.: Free Press, 1957.

Wright, James D., Peter H. Rossi, Sonia R. Wright, and Eleanor Weber-Burdin. *After the Clean-Up: Long-Range Effects of Natural Disasters.* Beverly Hills, Calif.: Sage, 1979.

Articles

Aaronson, Steve. "Predictions of Temblors Can Create New Hazards." *Science Digest,* Apr. 1978, 65–78.

Aggarwal, Yash P., and Lynn R. Sykes. "Earthquakes, Faults, and Nuclear Power Plants in Southern New York and Northern New Jersey." *Science,* 28 Apr. 1978, 425–29.

Aguirre, B. E. "The Long Term Effects of Major Natural Disasters on Marriage and Divorce: An Ecological Study." *Victimology* 5 (1980): 298–307.

Ahearn, Frederick L., Jr. "Disaster Mental Health: A Pre- and Post-Earthquake Comparison of Psychiatric Admission Rates." *Urban and Social Change Review* 14 (Winter 1981): 22–28.

Alexander, David. "Death and Injury in Earthquakes." *Disasters,* 1 Sept. 1985, 57–60.

Allen, Clarence R. "Earthquake Prediction: 1982 Overview." *Bulletin of the Seismological Society of America* 72 (Dec. 1982): S331–35.

Anderson, Don L. "Earthquakes and the Rotation of the Earth." *Science,* 4 Oct. 1974, 49–50.

"Animals Sense Change Before Quake Strikes." *Science Digest,* Mar. 1978, 18–19.

"Are U.S. Cities Ready for a Major Quake?" *U.S. News and World Report,* 3 Mar. 1980, 40.

"Arkansas: Shakes, Rattles, but No Roll." *Science News,* 1 Jan. 1983, 7.

"Assessing the Risk of Earthquakes in the Eastern United States." *Earthquake Information Bulletin,* Sept.-Oct. 1981, 175–78.

Bibliography

Atkinson, William. "Whole Lotta Shakin' Goin' On." *St. Louis Magazine*, Aug. 1982, 18–20.

Babb, Margaret E. "The Mansion House of Cahokia and Its Builder, Nicholas Jarrett." *Transactions of the Illinois State Historical Society* 31 (1924): 78–93.

Bader, Allan H. "A Guide to Earthquake Hazard Analysis." *Risk Management*, Feb. 1983, 22–30.

Barbina, G. "The Fruili Earthquake as an Agent of Social Change in a Rural Area." *Mass Emergencies* 4 (1979): 145–49.

Baum, Andrew, Raymond Fleming, and Laura M. Davidson. "Natural Disaster and Technological Catastrophe." *Environment and Behavior* 15 (May 1983): 333–54.

Bayard, Ralph. "The South Missouri Diocese in Embryo, 1824." *Bulletin of the Missouri Historical Society* 14 (Oct. 1957): 7–21.

Berry, Daniel. "The Illinois Earthquake of 1811 and 1812." *Transactions of the Illinois State Historical Society* 12 (1908): 74–78.

Bissell, Richard A. "Delayed-Impact Infectious Disease After a Natural Disaster." *Journal of Emergency Medicine* 1 (1983): 59–66.

Bolin, Robert, and Daniel J. Klenow. "Response of the Elderly to Disaster: An Age-Stratified Analysis." *International Journal of Aging and Human Development* 16 (1982-1983): 283–98.

Boyd, S. T. "Psychosocial Reactions of Disaster Victims." *South African Medical Journal* 60 (1981): 744–48.

Braile, Lawrence W., William J. Hinze, G. R. Keller, and Edward G. Lidiak. "The Northeastern Extension of the New Madrid Seismic Zone." In *Investigations of the New Madrid, Missouri, Earthquake Region*, Geological Survey Professional Paper no. 1236, edited by F. A. McKeown and L. C. Pakiser, 175–84. Washington: GPO, 1982.

Broadhead, Garland C. "The New Madrid Earthquake." *American Geologist* 30 (Aug. 1902): 76–87.

Brown, Stuart. "Old Kaskaskia Days and Ways." *Transactions of the Illinois State Historical Society* 10 (1906): 128–44.

Bufe, Charles G. "Earthquake Prediction: Latest Progress and Problems." *Earthquake Information Bulletin*, July-Aug. 1975, 3–6.

Burton, Paul W., Ian G. Main, and Roger E. Long. "Perceptible Earthquakes in the Central and Eastern United States (Examined Using Gumbel's Third Distribution of Extreme Values)." *Bulletin of the Seismological Society of America* 73 (Apr. 1983): 497–518.

"Buttressing Resists Quakes." *Engineering News Record,* 24 Jan. 1985, 84–85.

"CAD Assists in Earthquake Prediction." *Civil Engineering*, Feb. 1984, 18–20.

Campbell, Robert P. "Identify Your Vital Business Functions." *ABA Banking Journal,* Jan. 1985, 34–35.

Bibliography

"Can California Chimps Predict Earthquakes?" *New Scientist,* 4 Nov. 1976, 275.

Carey, John. "Geology Breaks New Ground." *Newsweek,* 19 Dec. 1983, 86–87.

"Chinese Earthquake: Latest in a Chain That Raises Worldwide Concern." *U.S. News and World Report,* 9 Aug. 1976, 18–23.

"Chinese-American Project Studies Earthquake Motion." *Physics Today,* Sept. 1981, 22.

Chinnery, Michael. "A Comparison of the Seismicity of Three Regions of the Eastern U.S." *Bulletin of the Seismological Society of America* 69 (June 1979): 757–71.

Clark, Blake. "America's Greatest Earthquake." *Reader's Digest,* Apr. 1969, 110–14.

"Coordinating Earthquake Monitoring." *Science News,* 21 Mar. 1981, 185.

Crabbs, Michael A., and Ken U. Black. "Job Change Following a Natural Disaster." *Vocational Guidance Quarterly* 32 (June 1984): 232–39.

Crabbs, Michael, and Edward Heffron. "Loss Associated with a Natural Disaster." *Personnel and Guidance Journal,* Feb. 1981, 378–82.

Derr, John S. "Earthquake Lights." *Earthquake Information Bulletin,* May-June 1977, 18–21.

———. "Earthquake Lights: A Review of Observations and Present Theories." *Bulletin of the Seismological Society of America* 63 (Dec. 1973): 2177–87.

———. "Locating the World's Earthquakes." *Earthquake Information Bulletin,* Sept.-Oct. 1974, 11–15.

———. "The National Earthquake Information Service." *Earthquake Information Bulletin,* July-Aug. 1977, 7–9.

Derr, John S., and Marvin A. Carlson. "Seismic Data Flow at NEIS." *Earthquake Information Bulletin,* July-Aug. 1977, 10–12.

"Determining the Depth of an Earthquake." *Earthquake Information Bulletin,* July-Aug. 1977, 16.

Diamond, Diana. "Quake and Shake." *Datamation,* 15 Nov. 1984, 80–95.

Douglas, John H. "Earthquake Research Part 2: Averting Disaster." *Science News,* 10 Feb. 1979, 90–92.

"Earthquake Hazard Reduction Program." *Science News,* 8 July 1978), 22.

"Earthquake History of Kentucky." *Earthquake Information Bulletin,* Jan.-Feb. 1973, 31–33.

"Earthquake Law." *Harper's New Monthly Magazine,* Sept. 1871, 580–84.

"Earthquake Mechanics in the Central United States." *Science,* 21 June 1974, 1285.

Bibliography

"An Earthquake or Two." *Harper's New Monthly Magazine,* Nov. 1855, 795–803.

"Earthquake Prediction: Modeling the Anomalous Vp/Vs Source Region." *Science,* Feb. 1975, 537–40.

"Earthquake Safety and Survival Rules." *Earthquake Information Bulletin,* May-June 1975, 19–20.

"Earthquakes and Resistivity." *Earthquake Information Bulletin,* May-June 1973, 10–11.

"Earthquakes and the Rotation of the Earth," *Science,* 4 Oct. 1974, 49–50.

"Earthquakes May Be Triggered by the Sun." *New Scientist,* 16 May 1968, 357.

"Earthquakes May Parallel, Not Cause, Polar Wobble." *New Scientist and Science Journal,* 21 Jan. 1971, 105.

"Earthquakes of the Western United States." *Atlantic Monthly,* Nov. 1869, 549–59.

"Earthquakes Strike a Light in the Sky." *New Scientist,* 24 June 1982, 846–47.

"Eastern Earthquakes." *Science Digest,* Aug. 1981, 25.

Edwards, J. G. "Psychiatric Aspects of Civilian Disasters." *British Medical Journal* 1 (1976): 944–47.

Ellson, Richard W., Jerome Milliman, and R. Blaine Roberts. "Measuring the Regional Economic Effects of Earthquakes and Earthquake Predictions." *Journal of Regional Science* 24 (1984): 559–79.

Erickson, K. T. "Loss of Communality at Buffalo Creek." *American Journal of Psychiatry* 133 (1976): 302–5.

Everndon, Jack F. "Earthquake Prediction: What We Have Learned and What We Should Do Now." *Bulletin of the Seismological Society of America* 72 (Dec. 1982): S343–49.

Finch, Ruy H. "The Missouri Earthquake of April 9, 1917." *Bulletin of the Seismological Society of America* 7 (Sept. 1917): 91–96.

Fisher, Mary Jane. "RAA Proposes Creation of Federal Earthquake Reinsurance Facility." *National Underwriter: Property and Casualty Edition,* 2 Feb. 1985, 1, 29.

Flamsteed, John. "A Letter Concerning the Natural Causes of Earthquakes." *Earthquake Information Bulletin,* July-Aug. 1975, 14–17.

Florman, Samuel C., "Disasters and Decision Making." *Technology Review,* Jan. 1984, 8–9.

"Forecast: Future Shock." *Time,* 24 Jan. 1977, 83.

"Forecasting of Earthquakes Shows Gains." *U.S. News and World Report,* 30 Apr. 1979, 65.

French, Steven P., and Mark S. Isaacson. "Applying Earthquake Risk Analysis Techniques to Land Use Planning." *APA Journal* 50 (Autumn 1984): 509–22.

Bibliography

Fukio, Yoshio, Kenji Kanjo, and Isao Nakamura. "Deep Seismic Zone as an Upper Mantle Reflector of Body Waves." *Nature* 272 (13 Apr. 1978): 606–8.

Fuller, Myron. "Our Greatest Earthquakes." *Popular Science Monthly,* July 1906, 76–86.

* "Getting Ready for a Big Quake." *Sunset.* Special supplement. Mar. 1982.

Gordon, David W., Theron J. Bennett, Robert B. Herrmann, and Albert M. Rogers. "The South-Central Illinois Earthquake of November 9, 1968: Macroseismic Studies." *Bulletin of the Seismological Society of America* 60 (June 1970): 953–71.

Gore, Rick. "Our Restless Planet Earth." *National Geographic,* Aug. 1985, 142–72.

Gribben, John. "Relation of Sunspot and Earthquake Activity." *Science,* 6 Aug. 1971, 558.

Gupta, Indra, and Otto W. Nuttli. "Spatial Attenuation of Intensities for Central U.S. Earthquakes." *Bulletin of the Seismological Society of America* 66 (June 1976): 743–51.

Gwynne, Peter. "Let's Make the Most of Our Faults." *National Wildlife,* Oct.-Nov. 1981, 34–38.

———. "Predicting Quakes." *Newsweek,* 3 Sept. 1979, 66.

Haggerty, Alfred G. "Earthquake Insurance Sales Nearing Limit." *National Underwriter: Property and Casualty Edition,* 29 June 1984, 1, 34.

———. "New Earthquake Policies Making a Big Hit." *National Underwriter: Property and Casualty Edition,* 12 August 1983, 34–35.

Hamilton, Robert M. "Quakes Along the Mississippi." *Natural History,* Aug. 1980, 70–75.

Hays, Walter. "The Design Earthquake and Earthquake Response Spectra." *Earthquake Information Bulletin,* July-Aug. 1974, 18–22.

Heinrich, Ross R. "Three Ozark Earthquakes." *Bulletin of the Seismological Society of America* 39 (Jan. 1949): 1–8.

Herman, Roger E. "Step-by-Step Guide to Disaster Preparedness." *American City and County,* Oct. 1982, 55–56.

Herrmann, Robert B., and Jose-Antonio Canas. "Focal Mechanism Studies in the New Madrid Seismic Zone." *Bulletin of the Seismological Society of America* 68 (Aug. 1978): 1095–1102.

Herrmann, Robert B., Shaing-Ho Cheng, and Otto W. Nuttli. "Archeoseismology Applied to the New Madrid Earthquakes of 1811–1812." *Bulletin of the Seismological Society of America* 68 (Dec. 1978): 1751–59.

Herrmann, Robert B., G. W. Fischer, and J. E. Zollweg. "The June 13, 1975 Earthquake and Its Relationship to the New Madrid Seismic

Zone." *Bulletin of the Seismological Society of America* 67 (Feb. 1977): 200–218.

Hill, David P. "The Sound of an Earthquake." *Earthquake Information Bulletin,* May-June 1976, 15–18.

Hill, Mary. "Earthquake Watch." *Earthquake Information Bulletin,* Mar.-Apr. 1976, 15–18.

Hinrichsen, Savillah T. "Pioneer Mothers of Illinois." *Transactions of the Illinois State Historical Society* 9 (1904): 505–13.

"How Deep Is an Earthquake?" *Science News,* 2 Mar. 1985, 140.

"How Earthquakes Affect Animals." *American Review of Reviews,* Jan. 1924, 103.

Howe, James R., and Thomas Thompson. "Tectonics, Sedimentation, and Hydrocarbon Potential of the Reelfoot Rift." *Oil and Gas Journal,* 12 Nov. 1984, 179–90.

Hunter, Roger N. "Earthquake Prediction: Fact and Fallacy." *Earthquake Information Bulletin,* Sept.-Oct. 1976, 24–25.

———. "Is There a Periodicity in the Occurrence of Earthquakes?" *Earthquake Information Bulletin,* Sept.-Oct. 1976, 4–7.

Hunter, Roger N., and John S. Derr. "Prediction Monitoring and Evaluation Program: A Progress Report." *Earthquake Information Bulletin,* May-June 1978, 93–95.

Hutton, Janice R., John H. Sorensen, and Dennis S. Mileti. "Earthquake Prediction and Public Reaction." In *Current Research in Earthquake Prediction,* vol. 1, edited by Tsuneji Rikitake. Dordrecht, Holland: Reidel, 1981.

Huyghe, Patrick. "Earthquakes: The Solar Connection." *Science Digest,* Oct. 1982, 73–75, 103.

"An Imperative Duty." *Center,* Sept.-Oct. 1983, 9–11.

"Is a Killer Earthquake Due to Hit the U.S.?" *U.S. News and World Report,* 14 Nov. 1983, 76.

Jackson, E. L. "Public Response to Earthquake Hazard." *California Geology,* Dec. 1977, 278–80.

———. "Response to Earthquake Hazard." *Environment and Behavior* 13 (July 1981): 387–416.

Johnston, Arch C. "A Major Earthquake Zone on the Mississippi." *Scientific American,* Apr. 1982, 60–68.

———. "A Natural Earthquake Laboratory in Arkansas." *Earthquake Information Bulletin,* Nov.-Dec. 1983, 210–15.

Johnston, William. "Sketches of the History of Stephenson County, Illinois." *Transactions of the Illinois State Historical Society* 30 (1923): 217–320.

Kanamori, Hiroo. "Are Earthquakes a Major Cause of the Chandler Wobble?" *Nature* 262 (22 July 1976): 254–55.

Bibliography

Kane, M. F., T. G. Hildenbrand, and J. D. Hendricks. "Model for the Tectonic Evolution of the Mississippi Embayment and Its Contemporary Seismicity." *Geology* 9 (Dec. 1981): 563–68.

Kenagy, G. J., and J. T. Enright. "Animal Behavior as a Predictor of Earthquakes? An Analysis of Rodent Activity Rhythms." *Zeitschrift für Tierpsychologie* 52 (1980): 269–84.

Kerr, Richard A. "Assessing the Risk of Eastern U.S. Earthquakes." *Science*, 9 Oct. 1981, 169–71.

———. "Earthquake Prediction Retracted." *Science*, 31 July 1981, 527.

———. "Earthquakes: Prediction Proving Elusive." *Science*, 28 Apr. 1978, 419–21.

———. "How to Catch an Earthquake." *Science*, 6 Jan. 1984, 38.

———. "Prospects for Earthquake Prediction Wane." *Science*, 2 Nov. 1979, 542–45.

———. "Quake Prediction by Animals Gaining Respect." *Science*, 16 May 1980, 695–96.

Kiecolt, K. Jill, and Joanne M. Nigg. "Mobility and Perceptions of a Hazardous Environment." *Environment and Behavior* 14 (Mar. 1982): 131–54.

King, Chi-Yu. "Electromagnetic Emissions Before Earthquakes." *Nature* 301 (3 Feb. 1983): 377.

———. "Radon: Emanation on the San Andreas Fault." *Earthquake Information Bulletin*, July-Aug. 1978, 136–38.

Koegler, R. R., and S. M. Hicks. "The Destruction of a Medical Center by Earthquake: Initial Effects on Patients and Staff." *California Medicine* 116 (1972): 63–69.

Lammlein, David R., Marc L. Sbar, and James Dorman. "A Microearthquake Reconnaissance of Southeastern Missouri and Western Tennessee." *Bulletin of the Seismological Society of America* 61 (Dec. 1971): 1705–16.

"The Lands of the Earthquakes." *Harper's New Monthly Magazine*, Mar. 1869, 480–81.

Landsberg, Helmut E. "Geophysical Predictions." *Earthquake Information Bulletin*, Nov.-Dec. 1978, 228–30.

Lang, Berel. "Earthquake Prediction: Testing the Ground." *Environmental Ethics* 5 (Spring 1983): 3–19.

Lee, W. H. K., and S. W. Stewart. "Microearthquake Networks and Earthquake Prediction." *Earthquake Information Bulletin*, Nov.-Dec. 1979, 192–95.

Legan, Marshall Scott. "Popular Reactions to the New Madrid Earthquakes, 1811–1812." *Filson Club Historical Quarterly* 50 (Jan. 1976): 60–71.

Lewis, Elias. "Earthquake Phenomena." *Popular Science Monthly*, Mar. 1873, 513–26.

Lifton, Robert Jay, and Eric Olson. "The Human Meaning of Total Disaster." *Psychiatry* 39 (Feb. 1976): 1–18.

Lindh, Allan. "Amplitude of Foreshocks as a Possible Precursor to Earthquakes." *Earthquake Information Bulletin,* Sept.-Oct. 1978, 162–64.

"Lineaments Linked with Earthquakes in Mississippi Embayment." *Earthquake Information Bulletin,* Jan.-Feb. 1976, 21–23.

Ling-huang, Shen. "Can Animals Help Predict Earthquakes?" *Earthquake Information Bulletin,* Nov.-Dec. 1978, 231–33.

Lissner, Will. "Coping with Catastrophe." *American Journal of Economics and Sociology,* July 1984, 299–300.

"Locating an Earthquake." *Earthquake Information Bulletin,* July-Aug. 1977, 14–15.

Logan, John M. "Animal Behaviour and Earthquake Prediction." *Nature* 265 (3 Feb. 1977): 404–5.

Lomnitz, Cinna, and Larissa Lomnitz. "Tangshan 1976: A Case History in Earthquake Prediction." *Nature* 271 (12 Jan. 1978): 109–11.

Londer, Randi. "After a Catastrophe." *Parade,* 8 Sept. 1985, 8.

"Looking Out for Luminous Phenomena." *Science News,* 24 Dec.-31 Dec. 1983, 412.

"Lunar Periodicity of Earthquakes." *Nature* 142 (9 July 1938): 81.

McGinnis, L. D., and C. Patrick Ervin. "Earthquakes and Block Tectonics in the Illinois Basin." *Geology,* Oct. 1974, 517–19.

McKeown, F. A., D. P. Russ, and P. C. Thenhaus. "Central Interior Seismic Source Zones: Summary of Workshop Convened June 10-11, 1980." In *Summary of Workshops Concerning Regional Seismic Source Zones of Parts of the Conterminous United States, Convened by the Geological Survey 1979–1980, Golden, Colo.,* Geological Survey Circular no. 898, edited by Paul C. Thenhaus 19–24. Washington: GPO, 1983.

McKibben, Wendy Lea. "Saving Systems from 'Shake, Rattle and Roll.'" *Infosystems,* May 1984, 66.

Magida, Phylis. "If Pandas Scream . . . an Earthquake Is Coming." *International Wildlife,* Sept. 1977, 37–39.

"Magnitude and Intensity: Measures of Earthquake Size and Severity." *Earthquake Information Bulletin,* Sept.-Oct. 1974, 21–28.

"Man-Made Earthquakes at Denver and Rangely, Colorado." *Earthquake Information Bulletin,* July-Aug. 1973, 4–9.

Mayes, Jewell. "Missouri History Not Found in Textbooks." *Missouri Historical Review,* Oct. 1927, 266–70.

"Measuring Success on the Richter Scale." *Nation's Business,* May 1983, 93.

Meiu, G., Angela Pricǎ and Cornelia Radu. "Catamnestic Aspects of Neurotic Relapses Following the Earthquake of March 1977." *Neurologie, Psihiatrie, Neurochirurgie* 26 (Oct.- Dec. 1981): 251–57.

Bibliography

"The Mental Effects of Earthquakes." *Popular Science Monthly*, June 1881, 257–60.

"Middle America's Fault." *Time*, 19 Nov. 1979, 66–67.

Mileti, Dennis S. "Public Perceptions of Seismic Hazards and Critical Facilities." *Bulletin of the Seismological Society of America* 72 (Dec. 1982): S13–18.

Mitchell, Brian J. "Radiation and Attenuation of Rayleigh Waves from the Southeastern Missouri Earthquake of October 21, 1965." *Journal of Geophysical Research* 78 (10 Feb. 1973): 886–99.

Mitchell, Brian J., C. C. Cheng, and W. Stauder. "A Three-Dimensional Velocity Model of the Lithosphere Beneath the New Madrid Seismic Zone." *Bulletin of the Seismological Society of America* 67 (Aug. 1977): 1061–74.

Mitchell, Brian J., and Braik Hashim. "Seismic Velocity Determinations in the New Madrid Seismic Zone: A New Method Using Local Earthquakes." *Bulletin of the Seismological Society of America* 67 (Apr. 1977): 413–24.

Mitchill, Samuel L. "A Detailed Narrative of the Earthquakes which occurred on the 16th day of December, 1811, and agitated the parts of North America that lie between the Atlantic Ocean and Louisiana; and also a particular account of the other quakings of the earth occasionally felt from that time to the 23rd and 30th of January, and the 7th and 16th of February, 1812, and subsequently to the 18th of December, 1813, and which shook the country from Detroit and the Lakes to New-Orleans and the Gulf of Mexico. Compiled chiefly at Washington, in the district of Columbia." *Transactions of the Literary and Philosophical Society of New York*, 1815, 281–307.

"Modified Mercalli Intensity Scale of 1931 (Adapted from Sieberg's Mercalli-Cancani Scale, Modified and Condensed)." *Earthquake Information Bulletin*, Jul.-Aug. 1977, 30–31.

Monagan, David. "How Animals Predict Earthquakes." *Science Digest*, June 1981, 92–95, 124.

Morrow, Lynn. "New Madrid and Its Hinterland: 1783–1826." *Bulletin of the Missouri Historical Society*, July 1980, 241–50.

Namel, Paul F., and William T. Ward. "Disaster Recovery Planning: Obligation or Opportunity?" *Risk Management*, May 1983, 44–47.

"Natural Hazards Information System Unveiled." *National Underwriter: Property and Casualty Insurance Edition*, 24 June 1983, 47.

Nelson, Wilbur A. "Reelfoot: An Earthquake Lake." *National Geographic*, Jan. 1924, 95–114.

"New Map Shows Rise and Fall of Earth's Crust." *Earthquake Information Bulletin*, Nov.-Dec. 1972, 17–17.

"A New Nightmare." *Economist*, 4 Sept. 1982): 91–92.

Newman, C. J. "Children of Disaster: Clinical Observations at Buffalo Creek." *American Journal of Psychiatry* 133 (1976): 306–12.

"The Next Big Earthquake Is on Its Way." *Changing Times,* Feb. 1977, 36–38.

Nichols, Ronald R., and June M. Buchanan-Banks. "Reducing Seismic Hazards: Rough Land-Use Planning." *Earthquake Information Bulletin,* Nov.-Dec. 1973, 3–9.

Nigg, Joanne M. "Communication Under Conditions of Uncertainty: Understanding Earthquake Forecasting." *Journal of Communication* 32 (Winter 1982): 27–36.

"NOAA Scientist 'Maps' Structural Responses to Earthquakes." *Earthquake Information Bulletin,* Mar.-Apr. 1973, 20–23.

"Nocturnal Earthquakes." *Nature* 239 (15 Sept. 1972): 131.

Nowak, Andrzej S., and Elizabeth L. M. Rose. "Seismic Risk for Commercial Buildings in Memphis." *Bulletin of the Seismological Society of America* 73 (Oct. 1983): 1435–50.

Nuttli, Otto W. "Damaging Earthquakes of the Central Mississippi Valley." In *Investigations of the New Madrid, Missouri, Earthquake Region,* Geological Survey Professional Paper no. 1236, edited by F. A. McKeown and L. C. Pakiser, 15-20. Washington: GPO, 1982.

―――. "Empirical Magnitude and Spectral Scaling Relations for Mid-Plate and Plate-Margin Earthquakes." *Technophysics* 93 (1983): 207–23.

―――. "Magnitude-Recurrence Relation for Central Mississippi Valley Earthquakes." *Bulletin of the Seismological Society of America* 64 (Aug. 1974): 1189–1207.

―――. "The Mississippi Valley Earthquakes of 1811 and 1812." *Earthquake Information Bulletin,* Mar.-Apr. 1974, 8–13.

―――. "The Mississippi Valley Earthquakes of 1811 and 1812: Intensities, Ground Motion, and Magnitudes." *Bulletin of the Seismological Society of America* 63 (Feb. 1973): 227–48.

Nuttli, Otto W., and James E. Zollweg. "The Relation Between Felt Area and Magnitude for Central United States Earthquakes." *Bulletin of the Seismological Society of America* 62 (Feb. 1974): 73–85.

O'Connell, Richard J., and Adam M. Dziewonski. "Excitation of the Chandler Wobble by Large Earthquakes." *Nature* 262 (22 July 1976): 259–62.

"Organization and Implementation of a Disaster Drill for Health Care Facilities." *National Safety News,* Mar. 1984, 65–68.

Owens, Thomas J., George Zandt, and Steven R. Taylor. "Seismic Evidence for an Ancient Rift Beneath the Cumberland Plateau, Tennessee: A Detailed Analysis of Broadband Teleseismic *P* Waveforms." *Journal of Geophysical Research* 89 (10 Sept. 1984): 7783–95.

Palm, Risa I. "Public Response to Earthquake Hazard Information." *An-*

nals of the Association of American Geographers 71 (Sept. 1981): 389–99.

Patton, Howard. "A Note on the Source Mechanism of the Southeastern Missouri Earthquake of October 21, 1965." *Journal of Geophysical Research* 81 (10 Mar. 1976): 1483–86.

Penick, James, Jr. "I Will Stamp on the Ground with My Foot and Shake Down Every House." *American Heritage* 27 (Dec. 1975): 82–87.

"Periodicity in Earthquakes." *Nature,* 20 Oct. 1934, 631.

Perkins, David M. "Seismic Risk Maps." *Earthquake Information Bulletin,* Nov.-Dec. 1974, 10–15.

Persinger, Michael. "Earthquake Activity and Antecedent UFO Report Numbers." *Perceptual and Motor Skills* 50 (1980): 791–97.

Person, Waverly J. "When an Earthquake Occurs. . . ." *Earthquake Information Bulletin,* July-Aug. 1977, 17–19.

Press, Frank. "Earthquake Prediction." *Scientific American* 232 (May 1975): 14–23.

————. "Mitigating Earthquakes: The Federal Role." *Earthquake Information Bulletin,* Nov.-Dec. 1977, 4–5.

Press, Frank, and Peter Briggs. "Chandler Wobble, Earthquakes, Rotation, and Geomagnetic Changes." *Nature* 256 (24 July 1975): 270–73.

"Progress on Quake Prediction." *Science News,* 11 Jan. 1975, 23.

"Prospects for Earthquake Prediction Wane." *Science,* 2 Nov. 1979, 543.

"Prospects for Short-Term Earthquake Prediction." *Science,* 4 Jan. 1985, 42.

Pusey, William Allen. "The New Madrid Earthquake: An Unpublished Contemporaneous Account." *Science,* 14 Mar. 1930, 285–86.

"Quake Plan Asks $366 Million." *Engineering News Record,* 17 Jan. 1985, 14–15.

"Quake Predictions: A Plan for 'What If?'" *Science News,* 15 Nov. 1975, 308–9.

"Radio Signals Before Quakes." *Science News,* 20 Mar. 1982, 200.

"Radon: A Message in Emissions?" *Science News,* 24 Sept. 1983, 206.

"Radon: Cause for Interest, Not Alarm." *Science News,* 21 Nov. 1981, 333.

"Radon Leaks Help to Predict Earthquakes." *New Scientist,* 27 Oct. 1983, 265.

Raleigh, Barry. "A Strategy for Short-Term Prediction of Earthquakes." *Bulletin of the Seismological Society of America* 72 (Dec. 1982): S337–42.

Raleigh, C. B. "Scientists' Responsibility for Public Information." *Science,* 31 July 1981, 499.

Raloff, J. "Big 'Shakes' Test Good Quake Design." *Science News,* 1 Sept. 1984, 134.

Rangell, L. "Discussion of the Buffalo Creek Disaster: The Course of Psychic Trauma." *American Journal of Psychiatry* 133 (1976): 313–16.

"Reading Seismic History in Tree Rings." *Science News,* 5 Feb. 1983, 90.

Reasenberg, Paul. "Unusual Animal Behavior Before Earthquakes." *Earthquake Information Bulletin,* Jan.-Feb. 1978, 42–49.

"Recent Events Shake Quake Prediction Theories." *Research and Development,* Sept. 1984, 83–85.

Repka, Janice. "Built to Take Quakes." *Mechanix Illustrated,* Sept. 1983, 80.

Rikitake, Tsuneji. "Anomalous Animal Behavior Preceding the 1973 Earthquake of Magnitude 7.0 That Occurred near Izu-Oshima Island, Japan." In *Current Research in Earthquake Prediction,* vol. 1, edited by Tsuneji Rikitake. Dordrecht, Holland: Reidei, 1981.

Robinson, Russell. "Earthquake Prediction: New Studies Yield Promising Results." *Earthquake Information Bulletin,* Mar.-Apr. 1974, 14–17.

Ross, Margaret. "The New Madrid Earthquake." *Arkansas Historical Quarterly* 27 (Summer 1968): 83–104.

Sampson, Francis A. "The New Madrid and Other Earthquakes of Missouri." *Missouri Historical Review,* July 1913, 179–99.

Schiller, Ronald. "Where Will the Next Earthquake Strike?" *Reader's Digest,* Oct. 1979, 120–24.

Schilt, F. Steve, and Robert E. Reilinger. "Evidence for Contemporary Vertical Fault Displacement from Precise Leveling Data Near the New Madrid Seismic Zone, Western Kentucky." *Bulletin of the Seismological Society of America* 71 (Dec. 1981): 1933–42.

"Scientists Closing In on Ability To Pinpoint Event." *Science Digest,* Mar. 1978, 12–17.

"Seismic Bearings Reduce Building Earthquake Damage." *Design News,* 9 Apr. 1984, 30–34.

"Seismic Reflections from the Deep Mantle." *Science,* 4 Jan. 1985, 42–43.

"Seismic Seiches." *Earthquake Information Bulletin,* Jan.-Feb. 1976, 13–15.

"Seismic Waves Yield New Deep Earth Data." *Earth Science* 34 (Fall 1981): 36–37.

"Seismographs: Keeping Track of Earthquakes." *Earthquake Information Bulletin,* July-Aug. 1977, 13.

"Sensing Quakes." *Time,* 29 Jan. 1979, 73.

"Shaking Up the Midwest." *Newsweek,* 25 Nov. 1968, 67.

Shaw, Evelyn. "Can Animals Anticipate Earthquakes?" *Natural History,* Nov. 1977, 14–20.

Bibliography

Shaw, John. "New Madrid Earthquake." *Missouri Historical Review*, Jan. 1912, 91–92.

Shepard, Edward M. "The New Madrid Earthquake." *Journal of Geology* 13 (Jan.-Feb. 1905): 45–62.

Simon, C. "Arkansas: Shakes, Rattles, but No Roll." *Science News*, 1 Jan. 1983, 7.

————. "Homing in on the Basin and Range." *Science News*, 14 May 1983, 308–9.

————. "Inner Geography." *Science News*, 30 Apr. 1983, 280–83.

————. "Quakes in the East." *Science News*, 10 Oct. 1981, 232–34.

Simon, Ruth B. "Animal Behavior and Earthquakes." *Earthquake Information Bulletin*, Nov.-Dec. 1975, 9–11.

Simon, Ruth B., and Carl Stover. Untitled article. *Earthquake Information Bulletin*, July-Aug. 1977, 24–29.

Sims, John H., and Duane D. Baumann. "Educational Programs and Human Response to Natural Hazards." *Environment and Behavior* 15 (Mar. 1983): 165–89.

Singh, Surendra. "Geomagnetic Activity and Microearthquakes." *Bulletin of the Seismological Society of America* 68 (Oct. 1978): 1533–35.

"Social Hazards of Earthquake Prediction." *Science News*, 8 Jan. 1977, 20–21.

Sorensen, John H. "Knowing How to Behave Under the Threat of Disaster: Can It Be Explained?" *Environment and Behavior* 15 (July 1983): 438–57.

"Southeastern U.S. Earthquake Study Begins." *Earthquake Information Bulletin*, Mar.-Apr. 1973, 5–7.

Spall, Henry. "Earthquakes and Plate Tectonics." *Earthquake Information Bulletin*, Nov.-Dec. 1977, 14–17.

————. "Introductory Remarks on NEIS From Lou Pakiser." *Earthquake Information Bulletin*, July.-Aug. 1977, 4–6.

————. "Paleomagnetism: One Key to Plate Tectonics." *Earthquake Information Bulletin*, Jan.-Feb, 1973, 13–17.

————. "Sociological Aspects of Earthquake Prediction." *Earthquake Information Bulletin*, May-June 1979, 102–4.

————. "Understanding Seismicity Within the Continents." *Earthquake Information Bulletin*, May-June 1979, 80–87.

————. "Water-Level Changes and Earthquake Prediction." *Earthquake Information Bulletin*, Jan.-Feb. 1978, 55–59.

Spence, William. "Measuring the Size of an Earthquake." *Earthquake Information Bulletin*, July-Aug. 1977, 21–23.

Stambolis, Costis. "Greek Physicists Perfect Earthquake Prediction." *New Scientist*, 12 May 1983, 359.

Stauder, William, and Gilbert Bollinger. "Pn Velocity and Other Seismic

Studies from the Data of Recent Southeast Missouri Earthquakes." *Bulletin of the Seismological Society of America* 53 (Apr. 1963): 661–79.

Stauder, William, Mark Kramer, Gerard Fischer, Stephen Schaefer, and Sean T. Morrissey. "Seismic Characteristics of Southeast Missouri as Indicated by a Regional Telemetered Microearthquake Array." *Bulletin of the Seismological Society of America* 66 (Dec. 1976): 1953–64.

Stauder, William, and Otto W. Nuttli. "Seismic Studies: South Central Illinois Earthquake of November 9, 1968." *Bulletin of the Seismological Society of America* 60 (June 1970): 973–81.

Stauder, William, and Andrew Pitt. "Note on an Aftershock Study, South-Central Illinois Earthquake of November 9, 1968." *Bulletin of the Seismological Society of America* 60 (June 1970): 983–86.

Steele, S. R., W. C. Hood, and J. L. Sexton. "Radon Emanation in the New Madrid Seismic Zone." In *Investigations of the New Madrid, Missouri, Earthquake Region,* Geological Survey Professional Paper no. 1236, edited by F. A. McKeown and L. C. Pakiser, 191–201. Washington: GPO, 1982.

Stierman, Donald J. "Earthquake Sounds and Animal Cues: Some Field Observations." *Bulletin of the Seismological Society of America* 70 (Apr. 1980): 639–43.

Still, Elmer G. "Volcanoes and the Sun and Moon." *Scientific American,* 21 June 1902, 443.

Street, R. L. "The Southern Illinois Earthquake of September 27, 1891." *Bulletin of the Seismological Society of America* 70 (June 1980): 915–20.

"Strong-Motion Instruments for VA Hospitals." *Earthquake Information Bulletin,* Nov.-Dec. 1972, 10–11.

Sugisaki, Ryuichi. "Deep-Seated Gas Emission Inducted by the Earth Tide: A Basic Observation for Geochemical Earthquake Prediction." *Science,* 12 June 1981, 1264–66.

Sullivan, Walter. "Quakes: U.S. Plans Ways to Cope." *Earthquake Information Bulletin,* Sept.-Oct. 1978, 160–61.

"Sun and Moon Tidal Forces May Trigger Some Classes of Earthquakes." *Earthquake Information Bulletin,* May-June 1975, 14–15.

Sunyari, Samuel W. "Computer Systems for Automatic Earthquake Detection." *Earthquake Information Bulletin,* May-June 1975, 17–21.

Taravella, Steve. "Future Shock?" *Business Insurance,* 2 July 1984, 3, 24.

Taylor, Verta. "Good News About Disaster." *Psychology Today,* Oct. 1977, 93–94, 124–26.

Toksoz, M. Nafi, Neal R. Goins, and C. H. Cheng. "Moonquakes: Mechanisms and Relation to Tidal Stresses." *Science,* May 1977, 979–81.

"Trees Tell Quake Tale." *Science Digest,* Aug. 1983, 29.

Bibliography

Turner, Ralph H. "The Challenge of Earthquake Prediction." *Earthquake Information Bulletin,* Jan.-Feb. 1978, 40–41.

———. "Media in Crisis: Blowing Hot and Cold." *Bulletin of the Seismological Society of America* 72 (Dec. 1982): S19–28.

von Hake, Carl A. "Earthquake History of Missouri." *Earthquake Information Bulletin,* May-June 1974, 24–26.

———. "Earthquake History of Tennessee." *Earthquake Information Bulletin,* Mar.-Apr. 1977, 37–39.

Ward, Peter. "Earthquake Prediction: A Perspective for a National Program." *Earthquake Information Bulletin,* Nov.-Dec. 1977, 10–13.

———. "Volunteers in the Earthquake Hazard Reduction Program." *Earthquake Information Bulletin,* July-Aug. 1978, 139–43.

Wellborn, Stanley N., and K. M. Chrysler. "Is a Killer Earthquake Due to Hit the U.S.?" *U.S. News and World Report,* 14 Nov. 1983, 74–76.

Wesson, Robert L., and John R. Filson. "The U.S. Earthquake Prediction Program." *Earthquake Information Bulletin,* Sept.-Oct. 1981, 164–74.

Westerfield, William U. "Developing Disaster Recovery Plans." *Chain Store Age Executive,* May 1983, 17.

White, Edgar. "Missouri History Not Found in Textbooks." *Missouri Historical Review,* Oct. 1925, 148–51.

Williams, Gurney, III. "How Science Forecasts the Killer Quakes." *Popular Mechanics,* Aug. 1981, 70–73, 102–6.

Williams, Richard L. "Science Tries To Break New Ground in Predicting Great Earthquakes." *Smithsonian,* July 1983, 41–50.

Winkler, Louis. "Catalog of U.S. Earthquakes Before the Year 1850." *Bulletin of the Seismological Society of America* 69 (Apr. 1979): 569–602.

Youd, T. Leslie, "Liquefaction." *Earthquake Information Bulletin,* Nov.-Dec. 1973, 11–17.

Zoback, Mark D., R. M. Hamilton, A. J. Crone, D. P. Russ, F. A. McKeown, and S. R. Brockman. "Recurrent Intraplate Tectonism in the New Madrid Seismic Zone." *Science,* 29 Aug. 1980, 971–76.

Zoback, Mark D., and Mary Lou Zoback. "States of Stress and Intraplate Earthquakes in the United States." *Science,* 3 July 1981, 96–104.

Public Documents

Ahearn, Frederic L., and Raquel E. Cohen, M.D., eds. *Disasters and Mental Health: An Annotated Bibliography.* Prepared for the Center for Mental Health Studies of Emergencies, National Institute of Mental Health. Rockville, Md.: National Institute of Mental Health, 1984.

Allen and Hoshall, Inc. *An Assessment of Damage and Casualties for Six Cities in the Central United States Resulting from Earthquakes in the New Madrid Seismic Zone.* Prepared for the Federal Emergency Management Agency. Washington: GPO, 1985.

Bodle, Ralph R. *United States Earthquakes: 1939.* Prepared for the U.S. Department of Commerce. Washington: GPO, 1941.

Brazee, Rutlage J., and William K. Cloud. *United States Earthquakes: 1956.* Prepared for the U.S. Department of Commerce. Washington: GPO, 1958.

———. *United States Earthquakes: 1958.* Prepared for the U.S. Department of Commerce. Washington: GPO, 1960.

Buntin, Wilbur R., Jr. *Report to Governor Martha Layne Collins.* Prepared for the Governor's Earthquake Hazards and Safety Technical Advisory Panel. Frankfort, Ky.: Division of Disaster and Emergency Services, 1985.

Central United States Earthquake Consortium. *The First Year: 1985.* N.p., 1985.

Coffman, Jerry L., and William K. Cloud. *United States Earthquakes: 1968.* Prepared for the U.S. Department of Commerce. Washington: GPO, 1970.

Coffman, Jerry L., Carl A. von Hake, and Carl W. Stover. *Earthquake History of the United States.* Rev. ed. Prepared for the U.S. Department of Commerce. Washington: GPO, 1982.

Federal Emergency Management Agency. *Behavior and Attitudes Under Crisis Conditions: Selected Issues and Findings.* Washington: GPO, 1984.

* ———. *Coping with Children's Reactions to Earthquakes and Other Disasters.* Washington: GPO, 1983.

* ———. *Earthquake Safety Checklist.* Washington: GPO, 1983.

* ———. *The Home Builder's Guide for Earthquake Design.* Washington: GPO, 1985.

———. *In Time of Emergency: A Citizen's Handbook.* Washington: GPO, 1985.

* ———. *Learning to Live in Earthquake Country: Preparedness for People With Disabilities.* Washington: GPO, 1984.

———. *National Earthquake Hazards Reduction Program: Fiscal Year 1984 Activities.* Washington: GPO, 1985.

———. Southern California Earthquake Preparedness Project. *Comprehensive Earthquake Preparedness Planning Guidelines: City.* Earthquake Hazards Reduction Series, no. 2. Washington: GPO, 1985.

———. *Comprehensive Earthquake Preparedness Planning Guidelines: Corporate.* Earthquake Hazards Reduction Series, no. 4. Washington: GPO, 1985.

Fowler, R. A. *Earthquake Prediction from Laser Surveying: A Report.*

Bibliography

Prepared for the National Aeronautics and Space Administration. Washington: GPO, 1968.

Fuller, Myron L. *The New Madrid Earthquake.* United States Geological Survey Bulletin no. 494. Washington: GPO, 1912.

Hays, Walter W. *Procedures for Estimating Earthquake Ground Motions.* Geological Survey Professional Paper no. 1114. Washington: GPO, 1980.

Hopper, Margaret G., S. T. Algermissen, and Ernest E. Dobrovolny. *Estimation of Earthquake Effects Associated with a Great Earthquake in the New Madrid Seismic Zone.* U.S. Geological Survey Open-File Report no. 83-179. Washington: GPO, 1983.

Illinois State Geological Survey. *Preliminary Proposal to DOSECC, Inc., for an Illinois Superdeep Drillhole.* Champaign: Illinois State Geological Survey, 1985.

Lander, James F., and William K. Cloud. *United States Earthquakes: 1962.* Prepared for the U.S. Department of Commerce. Washington: GPO, 1964.

McGinnis, Lyle D. *Earthquakes and Crustal Movement as Related to Water Load in the Mississippi Valley Region.* Illinois State Geological Survey Circular no. 344. Urbana: Illinois State Geological Survey, 1963.

McKeown, F. A., and L. C. Pakiser, eds. *Investigations of the New Madrid, Missouri, Earthquake Region.* Geological Survey Professional Paper no. 1236. Washington: GPO, 1982.

Murphy, Leonard M. *United States Earthquakes: 1954.* Prepared for the U.S. Department of Commerce. Washington: GPO, 1956.

———. *United States Earthquakes: 1955.* Prepared for the U.S. Department of Commerce. Washington: GPO, 1957.

Murphy, Leonard M., and Franklin P. Ulrich. *United States Earthquakes: 1949.* Prepared for the U.S. Department of Commerce. Washington: GPO, 1951.

Murphy, Leonard M., and William K. Cloud. *United States Earthquakes: 1953.* Prepared for the U.S. Department of Commerce. Washington: GPO, 1955.

Neumann, Frank. *United States Earthquakes: 1934.* Prepared for the U.S. Department of Commerce. Washington: GPO, 1936.

Nuttli, Otto W. *Evaluation of Past Studies and Identification of Needed Studies of the Effects of Major Earthquakes Occurring in the New Madrid Fault Zone.* Prepared for the Federal Emergency Management Agency. Washington: GPO, 1981.

Thenhaus, Paul C., ed. *Summary of Workshops Concerning Regional Seismic Source Zones of Parts of the Conterminous United States Convened by the U.S. Geological Survey 1979–1980, Golden, Colo.* Geological Survey Circular no. 898. Washington: GPO, 1983.

von Hake, Carl A., and William K. Cloud. *United States Earthquakes: 1963.* Prepared for the U.S. Department of Commerce. Washington: GPO, 1965.

————. *United States Earthquakes: 1965.* Prepared for the U.S. Department of Commerce. Washington: GPO, 1967.

————. *United States Earthquakes: 1967.* Prepared for the U.S. Department of Commerce. Washington: GPO, 1969.

U.S. Congress. House. Committee on Science and Technology. Subcommittee on Investigations and Oversight and Subcommittee on Science, Research, and Technology. *Earthquakes in the Eastern United States: Hearings.* 98th Cong., 2d sess., 1984. H. Doc. 141.

Unpublished Material

Steele, Sheila. "Spatial and Temporal Radon Variation in the New Madrid Seismic Zone." Master's thesis, Southern Illinois University at Carbondale, 1978.

Viitanen, Wayne John. "The Winter the Mississippi Ran Backwards: The Impact of the New Madrid, Missouri, Earthquake of 1811–1812 on Life and Letters in the Mississippi Valley." Ph.D. diss., Southern Illinois University at Carbondale, 1972.

Witherall, Charles R. "Seismic Analysis of Neely Hall." Master's thesis, Southern Illinois University at Carbondale, 1977.

Newspapers

Cairo Evening Citizen. 1934–.
Chicago Daily Tribune. 1895–.
Chicago Sun-Times. 1968–.
Daily Egyptian (Southern Illinois University at Carbondale). 1970–.
Memphis Commercial Appeal. 1968–.
New York Times. 1895–.
Southern Illinoisan. 1934–.
St. Louis Globe-Democrat. 1895–.
St. Louis Post-Dispatch. 1895–.
USA Today. 1983–.

Interviews

Ahearn, Frederick L., Jr. Telephone interview with author, 6 Dec. 1985.
Eidel, J. James. Telephone interview with author, 20 Dec. 1985.
Fischer, Hans. Interview with author, Carbondale, Ill., 13 Dec. 1985.
Flores, Paul. Telephone interview with author, 4 Dec. 1985.
Johnston, Arch. Telephone interview with author, 6 Dec. 1985.

Bibliography

Jones, E. Erie. Interview with author, Marion, Ill., 9 Dec. 1985.
Malinconico, Lawrence. Interview with author, Carbondale, Ill., 3 Dec. 1985.
McKeown, Frank A. Telephone interview with author, 5 Dec. 1985.
Mileti, Dennis. Telephone interview with author, 20 Dec. 1985.
Moy, Dr. Richard. Interview with author, Carbondale, Ill., 11 Dec. 1985.
Nuttli, Otto W. Telephone interview with author, 1982 and 4 Dec. 1985.
Quarantelli, E. L. Telephone interview with author, 10 Dec. 1985.
Schiff, Anshel J. Telephone interview with author, 4 Dec. 1985.

Index

Index

Index